Essential Quantum Mechanics for
Electrical Engineers

Essential Quantum Mechanics for Electrical Engineers

Peter Deák

WILEY-VCH

Author

Prof. Peter Deák
University of Bremen
Bremen Center for Computational
Materials Science
TAB-Gebäude
Am Fallturm 1
28359 Bremen
Germany

Cover
Background figure - "Qiwen"
(Shutterstock.com); diode - "coddie"
(Depositphotos.com)

Library of Congress Card No.: applied for

British Library Cataloguing-in-Publication Data
A catalogue record for this book is available from the British Library.

Bibliographic information published by theDeutsche Nationalbibliothek
The Deutsche Nationalbibliothek lists this publication in the Deutsche Nationalbibliografie; detailed bibliographic data are available on the Internet at <http://dnb.d-nb.de>.

Print ISBN: 978-3-527-41355-3
ePDF ISBN: 978-3-527-80581-5
ePub ISBN: 978-3-527-80583-9
Mobi ISBN: 978-3-527-80584-6

Cover Design Bluesea Design, McLeese Lake, Canada
Typesetting SPi Global Private Limited, Chennai, India
Printing and Binding Markono Print Media Pte Ltd, Singapore

Printed on acid-free paper

For my children and grandchildren.

Contents

Preface

My motivation for writing this book was the expected effect of nanotechnology on engineering, which will surely and significantly enhance the demand for the knowledge of quantum mechanics (QM). Although present-day micro- and optoelectronics can, to a degree, be understood using semiclassical models, this situation is going to change soon. The limits of development in the traditional (twentieth century) hardware have almost been reached. The upcoming devices – where switching happens at the level of single electrons, tunnel effects are actively utilized, and superposition states of electrons are used as qubits – are based on phenomena that cannot be grasped even approximately without the conceptual understanding of QM. Most students graduating in electrical engineering in the coming years will definitely be confronted in their professional career with the paradigm shift induced by the new technologies of the quantum world. This explains why the teaching of QM should begin early.

Although teaching QM to students of electrical engineering (and informatics) at the undergraduate level is becoming more and more widespread, there are hardly any textbooks written specifically for such courses. Typical books on QM are not well suited for engineers because of the excessive use of mathematics and because of the very abstract way of treatment with little or no applications relevant to them. QM books written for electrical engineers are usually either resorting to heuristics or aim at the band theory of solids (to be able to describe semiconductor applications), the latter being well beyond the possibilities provided by a bachelor curriculum. Based on my 25 years of experience in teaching QM for undergraduate students of electrical engineering and informatics, I have attempted to write a textbook, adjusted to their knowledge level and interests, which can be the basis of a two hours a week, one semester course.

From the viewpoint of electrical engineering, QM is primarily the physics of electrons. Its knowledge enables us to use them for information processing, storage, and display, as well as for lighting and energy production. Our organs of perception cannot register individual electrons, so we cannot really *imagine* what they are really like. As Richard Feynman has formulated it, the electron is not an object (what we can see or hold) but a concept, which can only be formulated mathematically. Accordingly, QM can only be formulated and interpreted mathematically, and this seems to be, at the first sight to undergraduates in electrical engineering, rather difficult to digest and of little practical interest. However, information technology is an important part of the trade, and the physics necessary to understand the hardware of electronic data processing and the conversion between electronic and electromagnetic information in data storage, transfer, and display has become indispensable. The majority of the graduates in electrical engineering and informatics will primarily be interested in system integration and algorithms, but optimal efficiency can only be achieved if they have an at least conceptual understanding about the working of the devices to be integrated and programmed. In addition, QM has changed our perception of reality very much, allowing a much deeper understanding of nature. Therefore, it should be part of the education of anybody striving for a bachelor degree in science and technology.

This book was specifically written for undergraduates of electrical engineering and shows the interlocking between the development of QM and the hardware of lighting technology, opto-, and microelectronics, as well as quantum information processing. I have attempted to demonstrate the surprising claims of basic QM in direct applications. The "Introduction" summarizes the basic concepts of classical physics and points out some of its failures, based on phenomena connected to lighting technology. These (blackbody radiation in the light bulb, emission spectrum of the gas fill, and cathode emission in discharge lamps) are analyzed in detail in Chapters 2–4, based on experiments which are famous in physics. It is shown that a surprising but rather controversial first explanation of the results could be provided in terms of the wave–particle duality principle. The use of that by Einstein led later to the discovery of the laser (which is also described). Chapter 5 goes beyond the duality principle and explains the particle concept of the QM and its consequences for electrical engineering (e.g., negative differential resistivity). Chapters 6–8 introduce the mathematical construction used for describing the state of a particle and to predict its properties. In Chapters 9 and 10, two examples of using this framework are shown (potential well and tunneling through a potential barrier), with applications, among others, in light-emitting diodes, infrared detectors, quantum cascade lasers, Zener diodes, and flash memories. The scanning tunneling microscope is, of course, explained and also the leakage currents in integrated circuits and the electric breakdown of insulators. Finally, in Chapters 11 and 12, some consequences of the QM for the chemical properties of atoms and for other many-electron systems (such as semiconductors) are depicted, giving also a brief insight into the potential hardware for quantum information processing. In Appendices A and B, the knowledge in classical physics and mathematics is summarized, which is a prerequisite to read the book. (It is strongly recommended to work through these appendices first.)

This book attempts to choose a middle course between abstract mathematics and applications. On the one hand, basic concepts and principles of the QM are introduced in the necessary mathematical formulation, but the mathematics is kept as simple as possible. Only those tools of advanced mathematics are used, which have to be learned in the electronic engineering curriculum anyhow, and even they are used to treat specific cases relevant for applications. Engineers usually prefer ready-made formulas over mathematical derivation. However, since the internal logic of QM is actually in the derivations, the most important ones are shown in this book – but only as footnotes. Chapters 9 and 10 are the two exceptions from this rule, where practically applicable formulas can be derived in elementary steps, helping the reader to gain a deeper understanding of specific cases. In addition, knowing very well that the targeted readers are mostly not too mathematics oriented, the book exploits the possibilities of multimedia: besides numerous figures and pictures, video clips and applets, accessible on the Internet, are used intensively. Application of QM often requires serious efforts with numerical calculations, but applets can ease the burden of that, allowing quick visualization of trends and easier cognition of graphically displayed information.

Finally, it should be noted that QM has raised many philosophical, epistemological questions. As far as possible, these have been swept under the carpet in this book, and – to use a philosophical term – a rather positivistic representation was chosen. Since this book was written for engineers, prediction of practical results should take precedence over philosophical interpretation. In addition, it is probably better to get a simplified but applicable picture, which later can be refined, than being bogged down right at the beginning with interpretational controversies.

I would like to express my gratitude to the people who have helped me to complete this book: Dr Bálint Aradi and Dr Michael Lorke, who have read and corrected the original German version, and Prof. Japie Engelbrecht who did the same with this English one.

Bremen 2016 *Peter Deák*

Owners of a printed copy can download color figures with the help of the QR code below

http://www.wiley-vch.de/de?option=com_eshop&view=product&isbn=9783527413553

1

Introduction: Classical Physics and the Physics of Information Technology

This chapter...
describes the view of classical physics about matter. The knowledge developed from these concepts has led to the first industrial revolution; however, it is not sufficient to explain many of the present technologies. The need for a substantial extension of physics is demonstrated by following the development of lighting technology.

1.1 The Perception of Matter in Classical Physics: Particles and Waves

The task of physics is the description of the state and motion of matter in a mathematical form, which allows quantitative predictions based on known initial conditions. Mathematical relationships are established for simplified and idealized model systems. Classical physics considers two basic forms of matter: *bodies* and *radiation*, characterized by mass m and energy E, respectively. The special relativity theory of Einstein (see Section A.9) has established that these two forms of matter can be mutually transformed into each other. In nuclear fusion or fission, for example, part of the initial mass will be converted into electromagnetic (EM) radiation (in the full spectral range from heat to X-rays), while energetic EM radiation can produce electron–positron pairs. The equivalence of mass and energy is expressed by $E = mc^2$. Still, the models used for the two forms of matter are quite different.

In classical physics, radiation is a *wave* in the ideally elastic continuum of the infinite EM field. Waves are characterized by their (angular) frequency ω and wave number k. These quantities are not independent, and the so-called dispersion relation between them, $\omega = \omega(k)$, determines the phase velocity v_f and group velocity v_g of the wave (see Sections A.5 and A.6). The energy of the wave is $E \sim v_f|\boldsymbol{E}_0|^2$, where $\boldsymbol{E}_0$ is the amplitude of the EM wave.

In contrast to the continuous EM field, bodies consist of discrete *particles*. The fundamental building blocks are the elementary particles[1] listed in Table 1.1.

1 Solids, fluids, and gases all consist of atoms with a nucleus and electrons. The nucleus consists of protons and neutrons, both of which are made up of quarks.

Essential Quantum Mechanics for Electrical Engineers, First Edition. Peter Deák.

Table 1.1 The elementary particles.

Particles	First generation	Second generation	Third generation
Quarks	Up (u)	Charm (C)	Top (t)
	Down (d)	Strange (S)	Bottom (b)
Leptons	Electron (e)	Muon (μ)	Tau (τ)
	e-Neutrino	μ-Neutrino	τ-Neutrino

The model of classical physics for particles is the *point mass*: a geometrical point (with no extension in space) containing all the mass of the particle. It has been found that the center of mass of an extended body is moving in such a way as if all the mass was carried by it, and all the forces were acting on it. Therefore, the concept of the point mass can even be applied for extended bodies. The point mass can be characterized by its position in space ($\boldsymbol{r}$) and by its velocity ($\boldsymbol{v}$), both of which can be accurately determined as functions of time. These kinematic quantities are then used to define the dynamic quantities, *momentum* $\boldsymbol{p}$, *angular momentum* $\boldsymbol{L}$, and *kinetic energy* T (see Section A.3).

The laws and equations of classical physics are formulated for point-mass-like particles and for waves in an infinite medium.

1.2 Axioms of Classical Physics

The motion of interacting point masses can be described by the help of the four Newtonian axioms (see Section A.2), which allow the writing down of an equation of motion for each point mass. Unfortunately, this system of equations can only be solved if the number of point masses is small or if we can assume that the distance between them is constant (rigid bodies). If the number of particles is high and the interaction between them is weak, a model of noninteracting particles (ideal gas) can be applied, and the system can be described by thermodynamic state variables. The changes in these are governed by the four laws of thermodynamics and by the equation of state. Actually, the state variables can be expressed by the Newtonian dynamic quantities, and the equation of state, as well as the four laws, can be derived from the Newtonian axioms with the help of the statistical physics and the kinetic gas theory (Figure 1.1).

The behavior of the EM field is described by the four axioms of Maxwell's field theory (see Section A.6). Far away from charges, these give rise to a wave equation, the solutions of which are the EM waves, traveling with the speed of light. The propagation of a local change in the field strength $\boldsymbol{E}$ can be given by the wave function $\boldsymbol{E}(\boldsymbol{r},t)$. The wave front is defined by the neighboring points in space where $\boldsymbol{E}$ has the same phase. Each point of the wave front is the source of a secondary elementary wave, and the superposition of the latter explains the well-known wave effects of refraction and diffraction.

Elastic and plastic (deformable) bodies (solids and fluids) contain a huge number of interacting particles, and neither the model of rigid bodies nor the model of

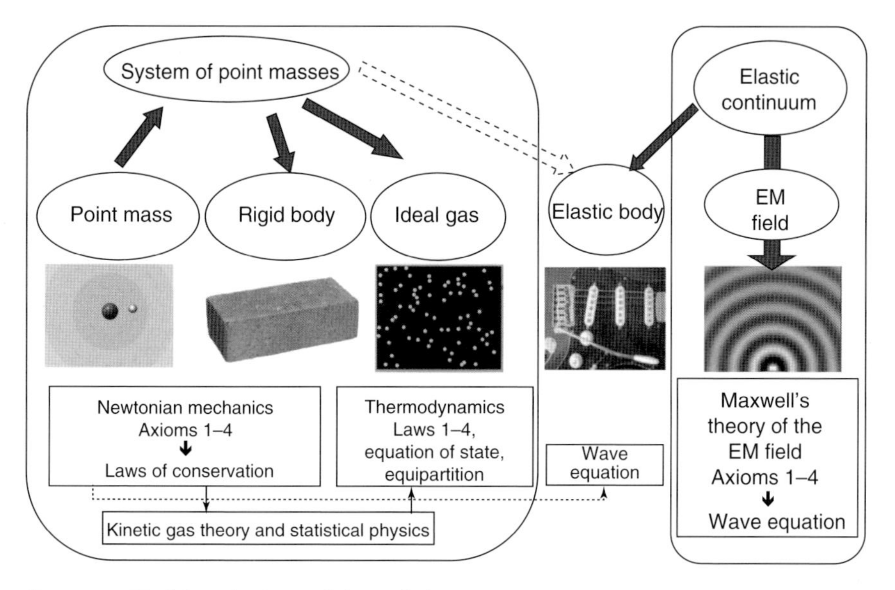

Figure 1.1 Models and axioms of classical physics.

the ideal gas can be applied. Instead, a continuum model can be used by assuming a continuous mass distribution, neglecting the corpuscular nature of the body. In the case of elastic bodies, the Newtonian equation for an infinitesimal volume of the continuum leads to a wave equation. The solutions are mechanical waves, corresponding to the propagation of local changes in the position (vibrations). The concepts and mathematics of mechanical and EM waves are quite similar.

1.3 Status and Effect of Classical Physics by the End of the Nineteenth Century

As we have seen in the previous subsection, classical physics contains two relatively independent parts: *mechanics* (from which also the thermodynamics can be derived) and *electrodynamics* (including optics).[2] Particles in classical mechanics are described by the concept of the point mass. In a conservative force field, where the potential energy $V(x,t)$ can be defined, the position of the point mass can be obtained from the Newtonian equation of motion:

$$m\ddot{x}(t) = \frac{dV(x,t)}{dx} \tag{1.1}$$

Historically (see Figure 1.2), the concept of the point mass has evolved, among others, from

2 More generally, one should talk about field theory. However, as we explain later, the only really relevant force field in electrical engineering is the EM field.

Figure 1.2 Scientists who were instrumental in the development of the point mass concept for particles. The pictures are taken from the public domain image collection of http://de.wikipedia.org

Figure 1.3 Application of mechanics and thermodynamics around the end of the nineteenth century. (a: The picture of the Eiffel tower was taken by the author. b: Power station, reproduced with permission of Daniel Hinze. c: Old locomotive, reproduced with permission of Herbert Schambach. d: The picture of the airplane was taken from the public domain of http://en.wikipedia.org.)

- the mathematical formulation of the observed regularities in the planetary motions (by, e.g., *J. Kepler*);
- the mathematical formulation of the experimentally observed motion of bodies on earth (by, e.g., *G. Gallilei*);
- the establishment of the axioms of mechanics (by *I. Newton*).

The concept of elementary particles, as the building blocks of a body, could later be confirmed in electrical measurements, too (e.g., by *E. Millikan,* who has shown that the electric charge of an oil droplet, floating in the field of a capacitor, can only be changed by discrete amounts, corresponding to the elementary charge, i.e., to the charge of an electron). The application of the principles of mechanics and thermodynamics around the end of the nineteenth century has led to the invention and optimization of structures and machines such as the ones shown in Figure 1.3.

The interpretation of light as a wave in the EM field is based on the wave equation, derived from Maxwell's axioms:

$$\frac{\partial^2 \psi(\mathbf{r}, t)}{\partial t^2} = v_f^2 \frac{\partial^2 \psi(\mathbf{r}, t)}{\partial \mathbf{r}^2} \tag{1.2}$$

where ψ is either the electric or the magnetic field and v_f the phase velocity (c in vacuum).

Historically (see Figure 1.4), the concept of the EM waves has evolved, among others, from

- the mathematical formulation of the laws of diffraction (e.g., by *A. J. Fresnel*);
- the mathematical formulation of the experimentally observed relations of electromagnetics (e.g., by *M. Faraday*);
- the establishment of the axioms of the EM field (by *J. C. Maxwell*).

The concept of the EM waves could later be confirmed in experiments (e.g., by *H. Hertz,* who could generate and detect radio waves), which are the basis of today's telecommunication technology. The application of the principles of electrodynamics (and wave optics) around the end of the nineteenth century has led to the invention of electrical lighting, the first forms of electrical data transfer and of "exotic rays" (see Figure 1.5).

Figure 1.4 Scientists who were instrumental in the development of the concept of electromagnetic waves. (The public domain of http://en.wikipedia.org.)

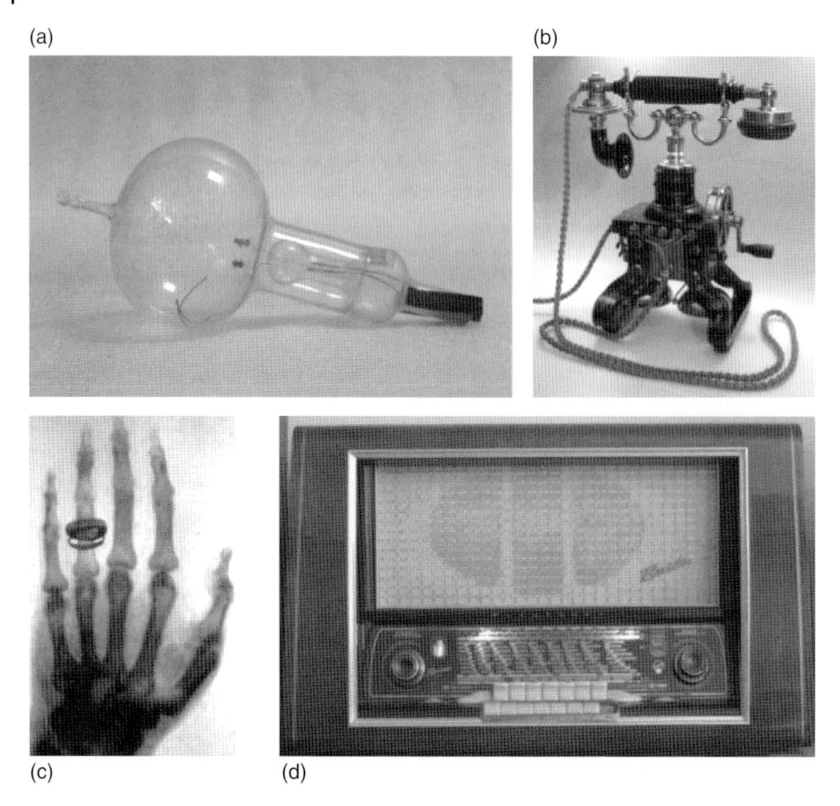

Figure 1.5 Application of electrodynamics and optics around the end of the nineteenth century. (a: The picture of the early light bulb by H. Ellgard was reproduced with permission of the Smithsonian National Museum Of American History. b: The picture of the old phone is by E. Etzold, https://en.wikipedia.org/wiki/Invention_of_the_telephone used under license: CC BY-SA 3.0 https://creativecommons.org/licenses/by-sa/3.0/. c: X-ray photograph, https://en.wikipedia.org/wiki/Wikipedia:Valued_picture_candidates/File:Anna_Berthe_Roentgen.gif Used under license: CC BY-SA 3.0 https://creativecommons.org/licenses/by-sa/3.0/. d: Old radio, https://en.wikipedia.org/wiki/Blaupunkt#/media/File:BlaupunktRadio1954.jpg used under license: CC BY-SA 2.0 DE https://creativecommons.org/licenses/by-sa/2.0/de/deed.en.)

1.4 Physics Background of the High-Tech Era

By the end of the nineteenth century, physics was considered a nearly completed discipline, with only a few remaining problems to be solved. As it happened, the latter required a revolutionary extension of the physical concepts. The inventions, following from that scientific revolution, have introduced the high-tech era of the late twentieth and the twenty-first century.

By the buzz word "information society," actually the exchange of electronically processed information through fast optical (EM) channels is meant.[3] Therefore, the devices interesting for electrical engineers today are semiconductor electronics for data processing and storage, optoelectronic elements to produce light or

3 Since a society is a structured group of individuals exchanging information, "information society," without further qualifiers is actually a tautology.

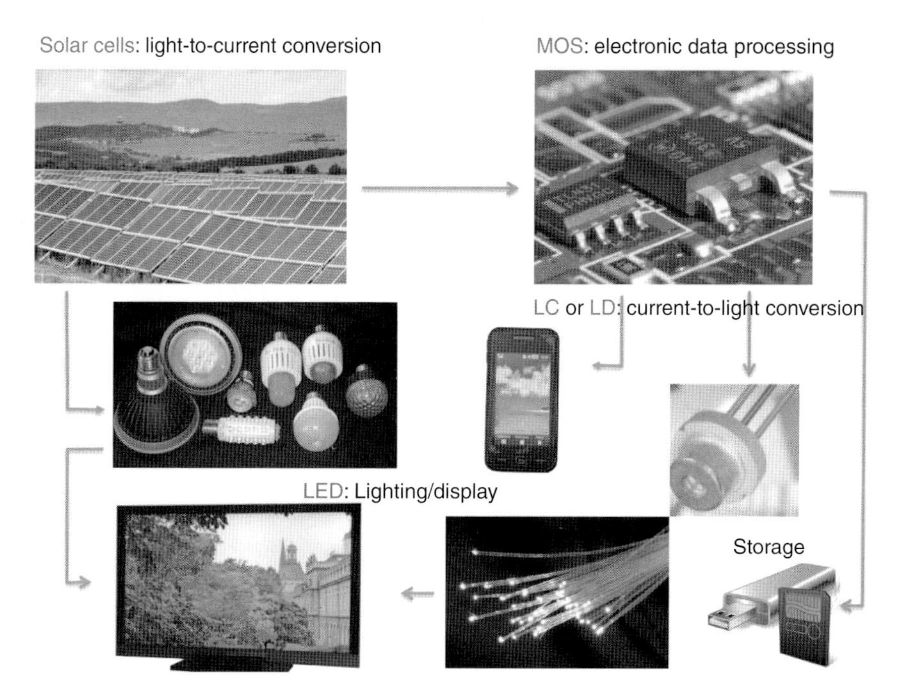

Figure 1.6 The hardware of information technology. Color online. (**solar cells** by Petr Kratochvil, http://www.publicdomainpictures.net/view-image.php?image=3061&picture=solar-power-plant Used under license: CC0 1.0 Universal https://creativecommons.org/publicdomain/zero/1.0/, **ICs** by Magnus Manske, https://commons.wikimedia.org/wiki/File:Chips_3_bg_102602.jpg Used under CC BY-SA 3.0 https://creativecommons.org/licenses/by-sa/3.0/, **blue laser diode (LD)**, Reproduced with permission of Visible Diode Lasers LLC, A Florida Corporation, **glass fiber** from https://de.wikipedia.org/wiki/Lichtwellenleiter#mediaviewer/Datei:Fibreoptic.jpg Used Under License: CC BY-SA.3.0 https://creativecommons.org/licenses/by-sa/3.0/, **LED lamps** by Geoffrey Landis, https://en.wikipedia.org/wiki/LED_lamp. Used Under License CC BY 3.0:https://creativecommons.org/licenses/by/3.0/, **smartphone & LED display** by the author, and the **SD memory** by Icons-land, Reproduced with permission of Icons Land.)

EM radiation, and displays (Figure 1.6). Lately in the latter, individual pixels are separately illuminated, and light for that is generated the same way as that for interior lighting, namely by light-emitting diodes (LEDs). Electronic information is converted into light modulation by laser diodes or oscillator (LC) circuits, while the reverse process is done by photodiodes. It should be noted that the operating principle of the photodiode is also utilized in most solar cells. The development of these devices has, on the one hand, pinpointed the limits of classical physics and, on the other hand, would not have been possible without its substantial extension.

1.5 Developments in Physics Reflected by the Development of Lighting Technology

The close connection between scientific and technical development can best be followed in lighting technology: from the light bulb, over the discharge lamps, to

Figure 1.7 Stages in the development of lighting technology: the light bulb, compact fluorescence lamp (discharge lamp), and the LED lamp.

the LED lamps (Figure 1.7). As described briefly later, every step of the technical development was coupled to a new chapter in the history of modern physics.

1.5.1 The Light Bulb (Incandescent Lamp)

The operating principles of an incandescent lamp are simple: current heats wire, wire radiates, and the gas fill keeps the wire stable and the bulb clear. The advantages are the low production costs and pollution level, as well as a natural spectral distribution of the radiated energy. The big disadvantage is the very low efficiency (< 20 lm/W, see Chapter 2), which has led to its ban in the European Union. It will be instructive to summarize our knowledge about the physics of the light bulb.

The electric current I in the wire of cross section A is carried by electrons with charge e, mass m, and velocity v:

$$I = envA \tag{1.3}$$

where n is the concentration (number/volume) of the electrons. The velocity of the electrons under constant voltage U in a wire of length l is

$$v = \frac{e\tau}{m}\frac{U}{l} \tag{1.4}$$

because of the scattering of electrons after the mean time τ (see also Section A.4.1). From Eqs (1.3) and (1.4) follows Ohm's law and the microscopic definition of the resistivity R,

$$I = \frac{e^2 n\tau}{m}\frac{A}{l}U = \frac{1}{R}U \tag{1.5}$$

The scattering amounts to friction which produces heat. The Joule heat produced in time t on the resistor R is

$$Q = I^2 Rt = UIt \tag{1.6}$$

Neglecting the gas fill and assuming that only the wire (with mass M and specific heat c_p) is heated, the temperature of the wire T is raised by

$$\Delta T = \frac{Q}{c_p M} \tag{1.7}$$

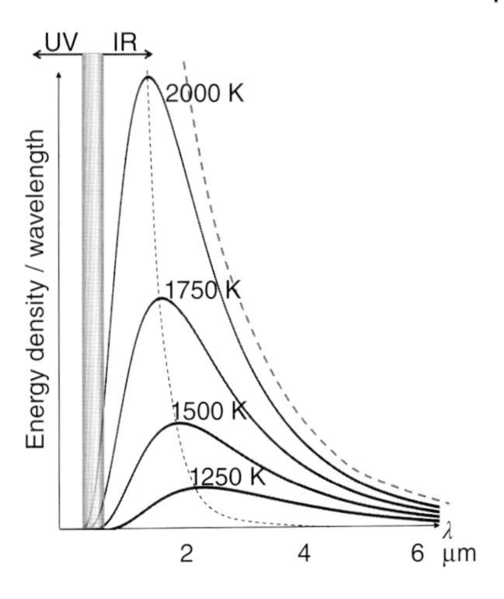

Figure 1.8 Spectral energy distribution of heat radiation. The black solid lines show (somewhat idealized) experimental curves, while the red dashed line shows the prediction of classical physics at $T = 2000$ K. A dotted line shows the shift of the wavelength with maximal energy as the temperature increases. Color online.

We know from experience that heated bodies (e.g., a ceramic stove) radiate heat and light of varying color, that is, a whole spectrum of EM waves. As we discuss in the next chapter, the equipartition theorem of classical physics leads to an energy distribution among different wavelengths λ, which is proportional to $1/\lambda^4$ (see dashed line in Figure 1.8). In contrast, spectroscopic measurements show an asymmetric curve, with a temperature-dependent maximum. The wavelength of the maximum determines the color, shifting from infrared toward ultraviolet with increasing temperature (solid lines in Figure 1.8). This phenomenon is the basis of the contactless temperature measurement by so-called *pyrometers* (see Chapter 2). The measured spectral distribution of the energy cannot at all be reproduced by classical physics. Obviously, if a simple tool, such as the light bulb, cannot be explained in the framework of a physical theory, we must go beyond it.

From Figure 1.8, it follows that pushing the maximum of the energy distribution into the visible range would require a very high temperature, about 6000 K,[4] obviously above the melting point of any metal. The refractory metal tungsten can be heated up to about 3000 K, and at that temperature only $\sim 5\%$ of the radiated energy is in the visible range. The rest only heats the environment. Understanding the laws of heat radiation could explain why this is so, and an accurate mathematical expression for the measured curves would help to design accurate pyrometers.

The explanation for the spectral distribution of the radiated energy, based on energy quanta as described in Chapter 2, was the first step toward quantum mechanics (QM).

1.5.2 The Fluorescent (Discharge) Lamp

The next stage in lighting technology, the discharge lamp (best known today in the form of the compact fluorescent lamp, CFL), is a lot more complicated than the light bulb. Let us consider a traditional fluorescent tube as shown in Figure 1.9.[5]

4 This is the temperature of the sun at its surface. Our organs of sight have developed to utilize the wavelength range radiated strongest by the sun.

5 The CFL is only smaller and wound up, made to fit in the socket of a light bulb.

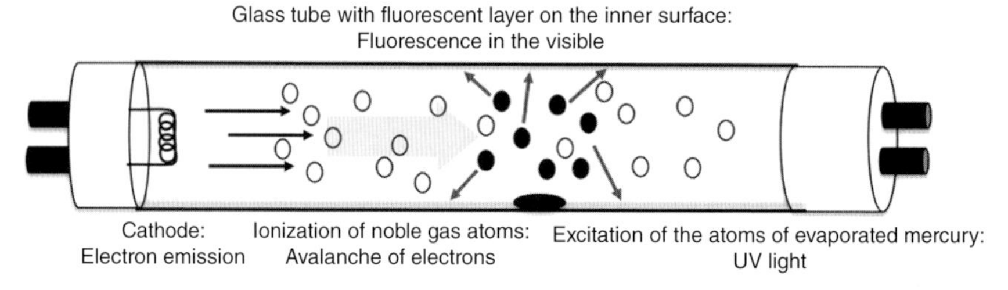

Figure 1.9 Schematic view of a fluorescent tube.

Fluorescent lamps are actually discharge tubes with two electrodes. The voltage between the heated tungsten electrodes induces electron emission and accelerates the electrons. Light atoms of the gas fill are ionized by the impact of the accelerated electrons, leading to an avalanche. The Joule heat produced by the current of the electrons evaporates a mercury droplet. Mercury atoms are also hit by the electrons but, instead of ionization, they are excited into a higher energy state. Returning to the ground state, the mercury atoms emit the excess energy as UV light. The latter excites the molecules of the fluorescent sheet[6] on the inside of the glass tube, making it to "fluoresce," that is, to emit white (or rather whitish) light. To produce UV light, which can be converted into white light by the fluorescent sheet, is very difficult without mercury. Lighter atoms, such as neon, xenon, krypton ("neon lamps"), or sodium, emit only one frequency (color) in the visible range.

This is a complicated system, indeed, with higher production costs and a lot more error sources than the light bulb. Still, the lifetime is about 10 times longer and the efficiency can be between 50 and 100 lm/W, that is, 2–5 times that of the light bulb. On the downside, though, the environmental load is much bigger, and the spectral distribution is ill fitted to the human eye. Optimization requires that we understand the physics of the individual processes; however, the limits of classical physics are quickly reached here too. For one, the voltage necessary to induce electron emission is much lower than that expected from the work function (the energy required for the electron to leave) of the electrode metal.[7] The very fact, however, that a given atom can only emit a few well-defined frequencies, runs completely against the expectation of classical physics (see Chapter 4).

The explanation provided for the spectral distribution of the heat radiation could have remained a curiosity, but to explain these discrete frequencies, again, the concept of quantization was needed. Such a model for the hydrogen atom has led to the questioning of the point-mass-like nature of the electron. Instead, wave-like behavior was attributed to it, and the investigation of that wave has led to the development of QM.

---

6 "Phosphor" in the language of the trade, although phosphor-containing substances are hardly ever used now.

7 The explanation for that is a phenomenon discovered in quantum mechanics, the tunnel effect: electrons can appear on the other side of a potential barrier without having to "jump" over it.

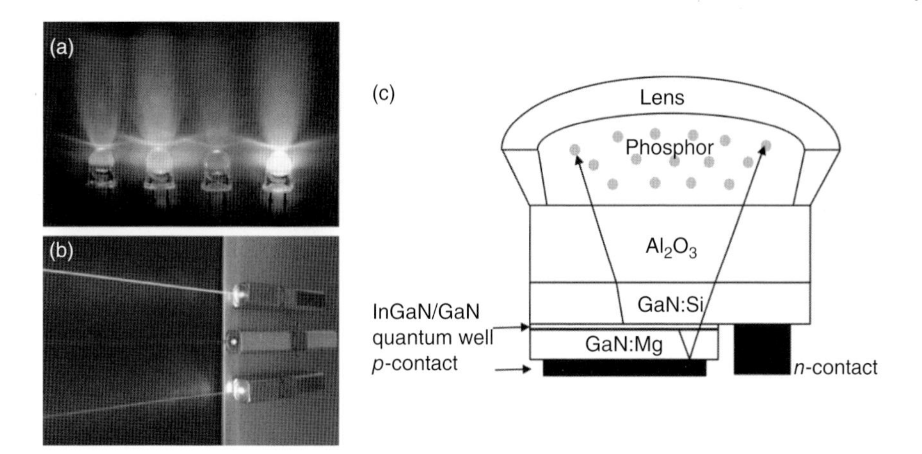

Figure 1.10 Light-emitting diodes and laser pointers of different colors, as well as the schematic (nonproportional) view of a white LED. Color online. (a: Reproduced with kind permission of Fraunhofer IAF. b: https://en.wikipedia.org/wiki/File:Lasers.jpeg#/media/File: Lasers.jpeg Used Under License: CC BY-SA 3.0 https://creativecommons.org/licenses/by-sa/3.0/.)

1.5.3 Light-Emitting and Laser Diodes

While the light bulb and the discharge lamps have evolved parallel with the development of the modern (quantum) physics and influenced that, the LEDs and diode lasers (Figure 1.10) were born as an application of this new physics. Electrical engineers are often using semiclassical models to describe electronic devices, such as diodes and transistors (even at the cost of having to assume an electron mass varying from material to material and sometimes even with voltage), but the development of functioning light-emitting and laser diodes (or, for that matter, solar cells) was only possible by understanding QM.

1.6 The Demand for Physics in Electrical Engineering and Informatics: Today and Tomorrow

The first computers, such as the legendary ENIAC (electronic numerical integrator and computer), were built of circuits with discharge tubes, but this is history by now. The basis of electronic data processing is solid-state (or semiconductor) electronics. The competition between discharge tubes and LEDs for illuminating pixels of a display and for interior lighting is not yet fully decided, but economic factors seem to make the victory of solid-state electronics very likely. Also considering solar cells and lithium batteries (with solid-state electrolytes), it is clear that for large areas of today's electrical engineering the relevant part of physics is the quantum-mechanics-based semiconductor physics. The latter will hopefully be described in a planned book, *Essential Semiconductor Physics for Electrical Engineers*. The QM needed there is described in this book.

The reason why QM is important for electrical engineers goes far beyond the need of semiconductor electronics. QM has changed our perception of matter considerably. The point mass, localized to a single geometrical point, and the infinite wave are both concepts idealized ad absurdum. As we will see, neither can be sustained as the model of a QM particle. Our cognition of the real world had to be expanded substantially. QM means a higher level of understanding and should be part of the world view of anyone aspiring for a B.Sc. in scientific–technical areas.

Beyond that, however, new concepts lead to new applications (such as the paradox of Schrödinger's cat to quantum information processing, or the massless, relativistic electrons of Dirac to ultrafast graphene electronics), which cannot be grasped even approximately without understanding QM. The billions spent on nanotechnology since the beginning of the twenty-first century will eventually come to bear fruit, and will, in the near future, change the work of an electrical engineer considerably. Electrical engineers are mostly interested in integrating devices into efficient and possibly programmable (automatized) systems. However, it is quite obvious that this is not possible without at least a conceptual understanding of the devices.[8] While semiclassical models could help engineers so far, this will certainly not be the case with the new nanoelectronic devices, where the wave nature of the electron is utilized in single-electron switches, or in quantum information processing and encrypting, where the information is coded, for example, into the magnetic properties of a single electron by light. The detailed physics of these future devices (e.g., quantum transport[9]) are, of course, beyond the scope of this book, which was written to accompany a B.Sc. course. The main aim here is to explain the basic concepts of QM, which are applied in today's devices such as *quantum well LEDs, cascade lasers,* or *flash memories.* Of course, understanding QM also requires some basic knowledge of classical physics and mathematics. Appendices A and B summarize that. In the following, we see the attempts for answering the questions, which we have raised in relation to the light bulb and the discharge lamp, on the basis of classical physics.

Summary in Short

- Classical physics knows two forms of matter: bodies, which are built of particles, and radiation, that is, waves in the EM field. Particles are treated as idealized point masses, and the EM field is considered as an infinite, ideal elastic continuum.
- It is assumed that the position and velocity of the point mass can, in principle, be accurately determined. The dynamic quantities, momentum, angular momentum, and kinetic energy (which have been defined originally by experiments) can then be unambiguously calculated with the help of the mass.
- The waves of the EM field are characterized by angular frequency and wave number, which determine the velocity of the wave. The energy of the wave is proportional to the phase velocity and to the absolute square of the amplitude.

---

8 Just one example from today's power electronics. Silicon carbide-based devices could diminish the energy loss by about 30%. However, unless the systems used for power switching are redesigned, they are far too expensive compared with silicon-based devices with higher loss.

9 For example, http://www.amazon.de/dp/0521631459/ref=rdr_ext_tmb.

- The classical concepts of point mass and waves allow quantitative predictions about the motion of particles and the EM field. They fail, however, if the two are in interaction, as is the case, for example, at the inter conversion of electron currents in light. The hardware of the information technologies (especially nanotechnologies) requires new concepts.

1.7 Questions and Exercises

The questions and exercises listed here refer in part to knowledge from Appendix A.

Problem 1.1 Which kinematic and dynamic quantities can be used to characterize the motion of a point mass?

Problem 1.2 How are canonically conjugate coordinates (q) and dynamic quantities (p) related to each other?

Problem 1.3 What is the basic assumption behind Ohm's law?

Problem 1.4 Demonstrate that a harmonic wave satisfies the wave equation!

Problem 1.5 What is the dispersion relation of light (i.e., of the EM field) in vacuum? Is the phase and the group velocity equal in this case?

Problem 1.6 What is stated in the equipartition theorem of Boltzmann?

2

Blackbody Radiation: The Physics of the Light Bulb and of the Pyrometer

In this chapter...
we are going to investigate the equilibrium between a heated body and the electromagnetic (EM) field, in order to obtain the wavelength (or frequency) distribution of the radiated energy, and its portion in the visible range. As we discuss here, to derive the correct distribution, an assumption is needed, which is in crass contradiction to the concepts of classical physics. Still, the assumption of quanta in the energy exchange between bodies and EM field by *Max Planck* has led to a quantitatively accurate reproduction of the observed spectra. Based on Planck's law of heat radiation, we can accurately measure the temperature of a hot body by a pyrometer and calculate the efficiency of the light bulb. These observations were the first indication of the serious problems and shortcomings of classical physics.

2.1 Electromagnetic Radiation of Heated Bodies

The light bulb or incandescent lamp is based on the phenomenon of EM radiation by heated bodies. Besides convection and conduction, this is the third form of heat transfer. The color of the heated body changes with the temperature T. Stars like the sun are very hot and appear to be yellowish white. The dotted line in Figure 2.1a shows the spectral distribution of solar radiation (that is, the radiation intensity as a function of wavelength) measured over the atmosphere.[1] On average, this spectrum is similar to the asymmetric curves shown in Figure 1.8, that is, to the spectrum of an ideal radiator with $T \approx 5800$ K (dashed line). As the solid line in Figure 2.1a shows, in a tungsten-halogen lamp with a wire temperature of about 3200 K, the wavelength λ_m, at which most of the energy is radiated, shifts to higher values. At temperatures below 3000 K, only a very small amount of the energy is radiated in the visible range (see. Figure 1.8), and the color of the heated body is red to yellow. According to the empirical law of *Wilhelm Wien*, $\lambda_m T$ = constant. The latter is the basis for contactless

1 The higher above the earth one goes, the lesser the radiation absorbed by gas molecules, but absorption in the atmosphere of the sun cannot be avoided.

Essential Quantum Mechanics for Electrical Engineers, First Edition. Peter Deák.

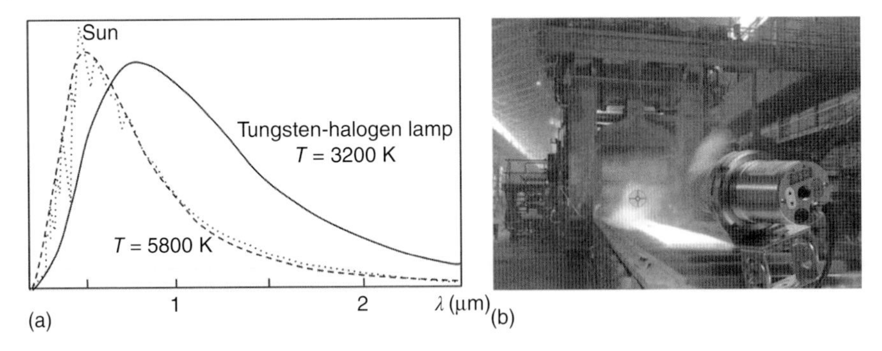

Figure 2.1 Spectrum of the sun and of a tungsten-halogen lamp (a) and contactless temperature measurement by a pyrometer (b). (b: Reproduced with permission of Kelly HCV Ltd.) Color online.

temperature measurement by a pyrometer, which can determine λ_m and, through that T, even for very hot bodies (Figure 2.1b). A good (accurate) pyrometer measures the intensity of the radiated light at two wavelengths.[2] From these two values, λ_m can be determined if the precise mathematical form for the ideal radiation curve is known.

The pyrometric method even allows measurement of the temperature of the stars, which was important for the development of cosmology, and that is why physicists have tried to deduce the law of radiation long before the invention of the light bulb. Their starting point was the assumption of thermal equilibrium between a heated body and the infinite EM field. This is a difficult task but one can utilize the empirical law of *Kirchhoff*: the capacity of a body for emission is the same as for absorption.[3]

Ideal radiators, such as the stars, can emit EM radiation of all frequencies, and the frequency distribution at the temperature of the sun appears to us as *white* light. Absorption of all frequencies means an *ideal blackbody,* which can easily be modeled by a metal box with a hole (Figure 2.2). EM rays of any frequency can enter through the hole but not leave easily. Through reflection on the internal

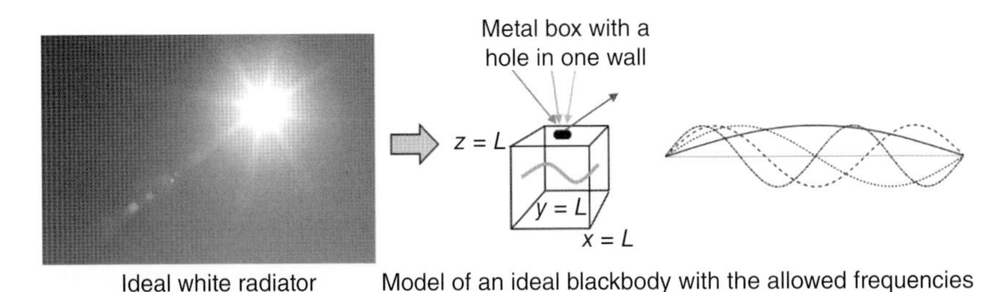

Figure 2.2 Modeling of an ideal radiator. Color online. (The picture of the sun is from rangizzz/fotolia.com, http://www.umweltbundesamt.de/themen/klima-energie/klimawandel/klima-treibhauseffekt.)

2 Cheap pyrometers measure at only one wavelength and use the Stefan–Boltzmann law (see Section 2.4) and material- and wavelength-specific emission coefficients.

3 Bodies can absorb the frequencies they can emit.

walls, standing EM waves will arise and a thermal equilibrium between the metal walls and the EM field inside the box can be achieved at any temperature T. The spectral energy distribution of the EM field in the box will be the same as that around an ideal radiator.

2.2 Electromagnetic Field in Equilibrium with the Walls of a Metal Box

The state of the EM field in the box can be derived from the Maxwell equations, which lead in this case (no charges in the box) to wave equations (see Eqs (A.25) and (A.26)). For the electric field

$$\frac{\partial^2 \mathbf{E}}{\partial t^2} = c^2 \Delta \mathbf{E} \tag{2.1}$$

The boundary condition is that the amplitude $\boldsymbol{E}_0$ should vanish at the walls. This is the same condition as in the 1D case of the guitar string (with both ends fixed), so the solutions are standing waves. In our 3D case, standing waves can arise in all three directions with two orthogonal transversal polarizations in each. According to Eq. (A.22), the allowed wave numbers in the three directions $i = x$, y, and z, are

$$k_{n_i} = \frac{n_i \pi}{L} \tag{2.2}$$

where n_i are arbitrary integers and L is the side length of the cubic box. From Eq. (2.2), with the help of the dispersion relation, Eq. (A.19), and with c as the velocity of light, it follows that the allowed frequencies of the standing EM waves in the box are

$$\omega_n = c|\mathbf{k}_n| = c\sqrt{k_{n_x}^2 + k_{n_y}^2 + k_{n_z}^2} \Rightarrow \nu_n = \frac{c}{2L}\sqrt{n_x^2 + n_y^2 + n_z^2} \tag{2.3}$$

Since every standing wave can be excited independently (similar to the case of plucking a guitar string), each allowed frequency ν_n represents a *thermodynamic degree of freedom* (in which energy can be stored). One can show[4] that the number of allowed frequencies between ν and $\nu + d\nu$ is

$$dZ = \frac{4\pi L^3}{c^3}\nu^2\, d\nu = \frac{4\pi V}{c^3}\nu^2\, d\nu \tag{2.4}$$

where $V = L^3$ is the volume of the box. If the time-average of the energy of an oscillator (standing wave) is $\langle\varepsilon\rangle$, then the spectral energy density (energy per unit volume and frequency) in the box is

$$\frac{1}{V}\frac{dE_\nu(T)}{d\nu} = 2\frac{\langle\varepsilon\rangle}{V}\frac{dZ}{d\nu} = \frac{8\pi}{c^3}\langle\varepsilon\rangle\nu^3 \tag{2.5}$$

where the factor 2 accounts for the two polarization directions.

4 The allowed frequencies must be positive and, up to a value of ν_n, the integers $n = (n_x, n_y, n_z)$ are in the positive eights of a sphere with radius $|n| = 2L\nu_n/c$ (c.f. Eq. (2.3)). The number Z of the allowed frequencies up to ν_n is equal to the volume:
$z(\nu_n) = \frac{1}{8}\frac{4\pi}{3}\left(\frac{2L}{c}\nu_n\right)^3 = \frac{4\pi L^3}{3c^3}\nu_n^3$ from which Eq. (2.4) follows after derivation with respect to ν_n.

2.3 Determination of the Average Energy per Degree of Freedom. Planck's Law

Boltzmann's equipartition theorem of classical physics assumes that the energy is equally distributed among the thermodynamic degrees of freedom, and at temperature T each has, on average, the energy

$$\langle \varepsilon \rangle_{\text{Boltzmann}} = \frac{1}{2}kT \tag{2.6}$$

where k is the Boltzmann constant. Substituting this into Eq. (2.5) leads to the so-called *Rayleigh–Jeans* law of radiation

$$\left[\frac{1}{V}\frac{dE_\nu(T)}{d\nu}\right]_{\text{Rayleigh–Jeans}} = \frac{8\pi}{c^3}kT\nu^2 \tag{2.7}$$

It is known, however, that the Boltzmann theorem leads to wrong results for the heat capacity of gases at normal temperatures (see A.7). As we discuss in the following, the equipartition fails here too, because Eq. (2.7) leads to the dashed line in Figure 1.8.

That is why *Max Planck* came up with a really surprising idea to reproduce the observed spectral distribution. He assumed that *energy can be exchanged between the EM field and a heated body* (in our case the wall of the metal box) *only by emission/absorption of quanta* $\Delta\varepsilon = h\nu$.[5] As a consequence, the energy of a standing EM wave with frequency ν_n can only be $\varepsilon_n = nh\nu$. With the help of Eqs (A.36) and (A.32), and using Eq. (B.3), we obtain the average energy per thermodynamic degree of freedom as

$$\langle \varepsilon \rangle_{\text{Planck}} = \frac{h\nu}{e^{h\nu/kT} - 1} \tag{2.8}$$

Using that in Eq. (2.5) leads to *Planck's law of radiation:*

$$\left[\frac{1}{V}\frac{dE_\nu(T)}{d\nu}\right]_{\text{Planck}} = \frac{8\pi}{c^3h^2}\frac{(h\nu)^3}{e^{h\nu/kT} - 1} \tag{2.9}$$

To convert Eqs (2.7) and (2.9) to wavelength dependence, the chain rule of derivation (Eq. (B14)) can be applied:

$$\frac{dE}{d\lambda} = \frac{dE}{d\nu}\left|\frac{d\nu}{d\lambda}\right| = \frac{dE}{d\nu}\left|\frac{d(c\lambda^{-1})}{d\lambda}\right| = \frac{dE}{d\nu}\frac{c}{\lambda^2} \tag{2.10}$$

So we obtain for the energy density distribution as a function of the wavelength the formulas

$$\left[\frac{1}{V}\frac{dE_\lambda(T)}{d\lambda}\right]_{\text{Rayleigh–Jeans}} = 4\pi\frac{kT}{\lambda^4} \tag{2.11}$$

$$\left[\frac{1}{V}\frac{dE_\lambda(T)}{d\lambda}\right]_{\text{Planck}} = \frac{8\pi kT}{\lambda^4}\frac{hc/\lambda kT}{e^{hc/\lambda kT} - 1} \tag{2.12}$$

which are depicted in Figure 2.3.

---

5 Planck's constant h is equal to $6.626 \cdot 10^{-34}$ Js.

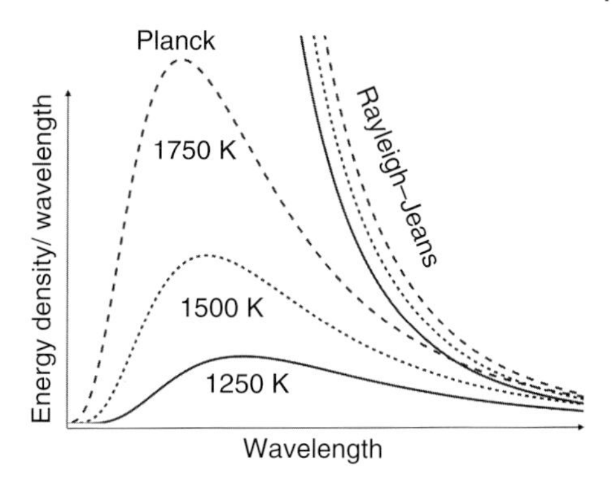

Figure 2.3 Spectral distribution of the energy density of the EM field in equilibrium with an ideal blackbody, according to Planck and according to Rayleigh and Jeans, at different temperatures.

In comparison with Figure 1.8, it is clear that the Rayleigh–Jeans formula is only valid for large wavelengths (small frequencies), while Planck's law of radiation describes the observed behavior in the entire wavelength range.

2.4 Practical Applications of Planck's Law for the Blackbody Radiation

We can apply our result for the spectral distribution to calculate the efficiency of the light bulb. Using Eq. (2.9), the total energy radiated at temperature T over the entire frequency range is [6]

$$\frac{E}{V} = \int_0^\infty \frac{1}{V}\frac{dE_\nu(T)}{d\nu}d\nu = \int_0^\infty \left[\frac{8\pi}{h^2c^3}\frac{(h\nu^3)}{e^{h\nu/kT}-1}\right]d\nu = \frac{8\pi^5k^4}{15h^3c^3}T^4$$
$$= 7.566 \times 10^{-16} T^4 (\mathrm{J\,m^{-3}}) \qquad (2.13)$$

that is, the total radiated energy is proportional to T^4, which is the empirically observed *Stefan–Boltzmann law* (used also in cheap pyrometers). In accurate pyrometers, T is adjusted in Eq. (2.12) to fit the light intensity measured at two wavelengths. The relation between T and λ_m with the maximum intensity can be determined from the condition that the derivative of Eq. (2.12) vanishes at the maximum. From that we obtain[7] *Wien's displacement law*

$$\lambda_m T = 0.2896\ (\mathrm{cm\ \ K}) \qquad (2.14)$$

It should be noted that the constants in the theoretically derived Eqs (2.13) and (2.14) agree quite accurately with the empirically determined values.

To calculate the efficiency of an incandescent lamp, one should consider the sensitivity of the human eye to various frequencies. Figure 2.4 shows the

---

6 The integral can be brought into a standard form by substituting $x = h\nu/(kT)$ and $dx = h\,d\nu/(kT)$, and

$$E = \frac{8\pi k^4}{h^3c^3}T^4 \underbrace{\int_0^\infty \frac{x^3}{e^x-1}dx}_{\pi^4/15} = \frac{8\pi^5k^4}{15h^3c^3}T^4$$

7 The condition $\frac{d}{d\lambda}\left[\frac{1}{V}\frac{dE_\lambda(T)}{d\lambda}\right] = \frac{d}{d\lambda}\left[\frac{8\pi hc}{\lambda^5}\frac{1}{e^{hc/\lambda kT}-1}\right] = \frac{8\pi hc}{\lambda^6}\frac{1}{(e^{hc/\lambda kT}-1)^2}\left[5 - e^{\frac{hc}{\lambda kT}}\left(5 - \frac{hc}{\lambda kT}\right)\right] = 0$ is satisfied if the expression in the last square bracket is zero. Numerical solution of the problem results in Eq. (2.14).

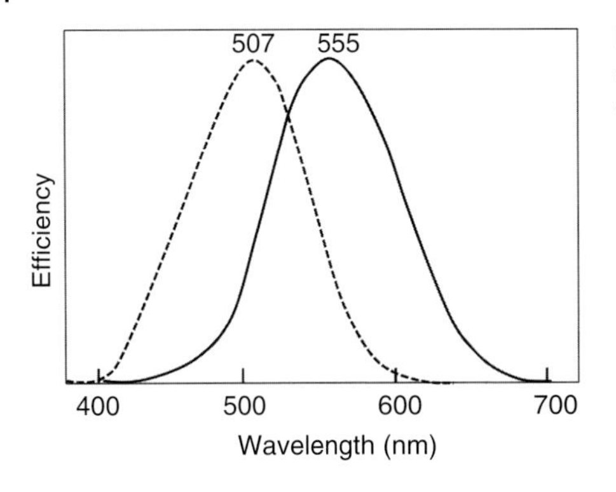

Figure 2.4 Phototopic (solid line) and scotopic (dashed line) spectral luminous efficiency, $v(\lambda)$.

measured spectral luminous efficiency of the eye, $v(\lambda)$, at day (phototopic) and at night (scotopic).

The *effective radiated power* (or *luminous flux*) of a light source is the energy, which is useful for the eye, and is radiated in Δt time from the whole A surface in all directions (in the whole range of the solid angle Ω):

$$\Phi = \frac{1}{\Delta t}\int\int \left(\frac{1}{V}\frac{dE}{d\lambda}\right) v(\lambda) d\Omega\, dA \tag{2.15}$$

where the spectral energy density distribution has to be taken from Eq. (2.12) and $v(\lambda)$ is usually approximated by a Gaussian at 555 nm. The luminous flux has its own SI unit: 1 lumen (lm) is equal to 1/683 watt (W). Its relation to the invested electric power $P_{in} = UI$ defines the efficiency of the light source:

$$\eta = \frac{\Phi(\text{lumen})}{P_{in}(\text{watt})} \tag{2.16}$$

As shown in Table 2.1, the efficiency of the incandescent lamps are less than 20 lm W^{-1} (i.e., about 3%), which is already near the theoretical limit, and could not have been improved in 10 years. That was the reason for their ban in the EU at the beginning of 2012.

It should be noted, though, that the common light bulb is not only less expensive but a lot more "eye-friendly" than the fluorescent lamps, and its waste is also less problematic. Despite its simplicity, it is by no means a "low-tech" product either. It represents metallurgy at its peak: the wire, wound up in a double helix, is up to 10 m long with a diameter of ~ 50 μm. The tolerance in the latter must be better than ±2 μm, to avoid local increase in the resistance, which could lead to fusing due to the excess Joule heat ("hot spot"). A lot of

Table 2.1 Characteristics of the incandescent lamps.

	Power consumption (W)	Electric loss (%)	Efficiency 2000 (lm W^{-1})	Efficiency 2010 (lm W^{-1})	Lifetime (h)
Common light bulb	15–150	0	9–15	9–15	1000
Halogen lamp	25–500	5–20	11–18	16–18	3000

development work was also invested in choosing the wire material – from the original carbon fibers (Edison's General Electric), through tungsten (by the late Hungarian company TUNGSRAM), to tungsten-based alloys (OSRAM) – and the gas fill – from vacuum (to prevent burning of the wire), through nitrogen–argon mixtures (to prevent the deposition of evaporated tungsten on the inner wall of the bulb), to the halogen lamps (where evaporated tungsten is carried back to the wire by tungsten halogenide molecules).

2.5 Significance of Planck's Law for the Physics

The accuracy by which the constants of Wien's law and of the Stefan–Boltzmann law could be theoretically derived was a convincing argument for Planck's hypothesis. However, as Planck himself has emphasized, this assumption does not follow from any of the axioms of classical physics and, in fact, runs against its concepts. Emission and absorption of quanta means that the energy of the EM field, as well as of a body (here the metal wall), can only be increased in discrete steps and not continuously. The classical definition of the energy (stored work, cf. Eq. (A.8)) means, however, a continuous quantity. Furthermore, Planck's hypothesis connects the energy to frequency (color), while in classical physics the energy of radiation depends on the phase velocity and the field amplitude (cf. Eq. (A.27)). However, the stepwise increase of the heat capacity in gases (cf. Figure A.5) confirms the quantum hypothesis. This means that the quantization of the energy is a physical reality, and classical physics needs to be significantly extended to be able to explain that. The quantum mechanics and, with that, the second technical revolution that resulted in today's information society, started with Max Planck.

Finally, it should be noted that Planck's hypothesis means, of course, that the temperature can also be increased stepwise only. However, these "steps" are very small. For example, absorption of an energy quantum corresponding to red light with a wavelength of 700 nm

$$h\nu_{\text{red}} = 6.6 \times 10^{-34}\ (\text{J s})\ \frac{3 \times 10^{8}\ (\text{m/s})}{700 \times 10^{-9}\ (\text{m})}$$
$$= 2.8 \times 10^{-19}\ (\text{J}) = 6.8 \times 10^{-23}\ (\text{kcal}) = 1.8\ (\text{eV})$$

increases the temperature of 1 liter of water by $6.7{\times}10^{-23}$ K only. In other words, the stepwise increase hardly differs from being continuous. Still, the quantum hypothesis is unavoidable for explaining the blackbody radiation. One should think of the quantization as a walk on sand. For the macroscopic foot, it appears to be smooth but through a microscope it clearly consists of fine grains.

Summary in Short

- A heated body radiates EM waves with a spectral distribution that has a temperature-dependent maximum. According to Kirchhoff's law, the emitted frequencies can also be absorbed. The spectral energy distribution in the EM field can be derived, instead of assuming an ideal (white) radiator, an ideal blackbody instead. Through emission and absorption, a thermal equilibrium should exist between the EM field and the body at any temperature T.

- A tractable model for the ideal blackbody is a cubic metal box with a small hole in one of the walls. The state of the EM field in the box can be obtained from Maxwell's wave equation, taking into account the boundary conditions. The result is a superposition of standing waves with frequencies $\nu_n = \frac{c}{2L}\sqrt{n_x^2 + n_y^2 + n_z^2}$ (with n_i being integers, c the light velocity, and L the edge length of the box).
- The number of allowed frequencies ν_n between ν and $\nu + d\nu$ is

$$dZ = \frac{4\pi L^3}{c^3}\nu^2 d\nu$$

Each of these frequencies corresponds to a thermodynamic degree of freedom with an average energy of $\langle\varepsilon\rangle$, and the energy density in the field at frequency ν is $\frac{1}{V}\frac{dE(\nu)}{d\nu} = \frac{1}{L^3}\langle\varepsilon\rangle\frac{dZ}{d\nu} = \frac{8\pi}{c^3}\nu^2\langle\varepsilon\rangle$.
- According to classical physics, each thermodynamic degree of freedom should have, on average, the energy $\langle\varepsilon\rangle = \frac{1}{2}kT$, from which $\frac{1}{V}\frac{dE(\lambda)}{d\lambda} \sim \frac{kT}{\lambda^4}$ follows. This, so-called Rayleigh–Jeans law of radiation reproduces, however, the observed spectra only for large wavelengths.
- According to Planck's hypothesis, after raising the temperature of a body, the thermal equilibrium with the EM field can only be restored by emission of energy quanta $h\nu$. This means that the energy of the individual thermodynamic degrees of freedom is discrete, $\varepsilon_n = nh\nu$. The average energy is, therefore,

$$\langle\varepsilon\rangle = \frac{h\nu}{e^{h\nu/kT} - 1}$$

and from that follows Planck's law of heat radiation:

$$\frac{1}{V}\frac{dE_\lambda(T)}{d\lambda} \sim \frac{kT}{\lambda^4}\frac{hc/\lambda kT}{e^{hc/\lambda kT} - 1}$$

This reproduces the observations accurately for the entire wavelength range.
- SIGNIFICANCE OF PLANCK'S HYPOTHESIS: In the interaction of standing EM waves and oscillating atoms of a body, the exchange of energy can only occur in quanta. A stepwise increase of energy cannot be explained by classical physics, questioning its validity!
- PRACTICAL CONSEQUENCES: Determination of the constants in Wien's displacement law, $\lambda T = 0.2896$ (cm K) and in the Stefan–Boltzmann law of heat radiation, $E = \frac{8\pi^5 Vk^4}{15h^3c^3}T^4$.

2.6 Questions and Exercises

Problem 2.1 By what factor should the electric power be increased to raise the temperature of the wire in a light bulb from 2000 to 2500 °C? Let us assume that the wire is heated by 100% of the electrical energy and can be regarded as ideal radiator, following the law of Stefan and Boltzmann.

Problem 2.2 The surface temperature of the star Aldebaran is 4100 °C. Estimate the wavelength λ_m at which the most energy is radiated. What is the color of Aldebaran?

Problem 2.3 Explain why Planck's law is in contradiction with the concepts of classical physics.

3

Photons: The Physics of Lasers

In this chapter...

The experimentally confirmed radiation law of Planck was derived with the assumption that, in the interaction between a heated body and the electromagnetic (EM) field, energy can be exchanged only in quanta. In case of the metal box, with which we have modeled the ideal blackbody, this condition means that the energy of the standing EM waves in the box can have only discrete values, $\varepsilon_n = nh\nu$ (where ν is the frequency of the standing wave, h is the Planck's constant, and $n = 0, 1, 2, \ldots$). This chapter shows how the investigation of other cases of interaction between the EM field (UV radiation) and bodies (electrons) has led to the extension of the energy quantization to freely propagating EM waves and to introducing the concept of the light particle (photon). While *Albert Einstein* tried to derive Planck's law with the help of photons, he discovered a physical effect (stimulated emission of photons), which has led later to the development of the laser. Lasers are now ubiquitous tools in technology, from optical data transmission, and optical measurements to processing of metals and to medicine. Therefore, both the basic principles and the most important realizations are explained here briefly.

3.1 The Photoelectric Effect

The photoelectric effect, discovered by *Heinrich Hertz*, is the emission of electrons from a metal surface upon UV irradiation. It can be explained by assuming that the EM wave, $\mathbf{E} = \mathbf{E}_0 e^{i(\omega t - \mathbf{kr})}$, provides energy for the electrons so they can overcome the attraction of the nuclei.[1] The energy transferred by the wave to a unit surface in unit of time is the intensity I. Therefore, a unit surface of the metal receives in Δt time the energy $E = I\Delta t = (1/2)\varepsilon_0 c|\mathbf{E}_0|^2\Delta t$ (where ε_0 is the electric constant, see Eq. (A.27)). One expects that the kinetic energy of the electrons, $T = \frac{1}{2}mv^2$, should be proportional to the intensity I of the UV light.

The photoelectric effect can be investigated in a cathode ray tube (CRT), depicted schematically in Figure 3.1. Similar to the discharge lamps, the CRT

1 Their increased kinetic energy exceeds the potential energy that binds them in the metal.

Essential Quantum Mechanics for Electrical Engineers, First Edition. Peter Deák.

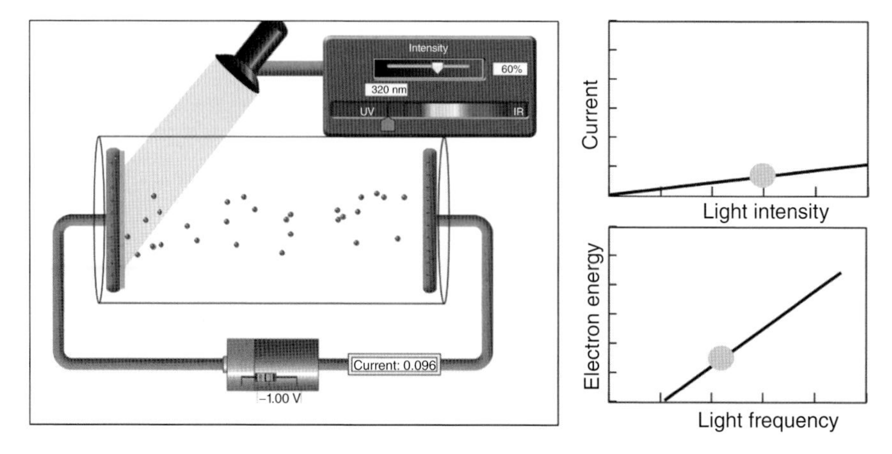

Figure 3.1 The photoelectric effect. Color online. (http://phet.colorado.edu/en/simulation/photoelectric.)

contains a cathode and an anode,[2] but without a gas fill (i.e., it is under vacuum). The electrons, emitted from the cathode, establish a current density $j = env$, where v is the velocity, n the concentration (number density), and e the charge of the electrons (see Eq. (1.3)). Detailed CRT experiments to investigate the photoelectric effect were carried out by *Philip von Lenard.* He observed that electrons were collected by the anode almost instantaneously after turning on the UV source, that is, in a time shorter than he could measure. Let us consider that, using a UV lamp of $P = 60$ W power consumption and $\eta \approx 10\%$ efficiency at a distance of $D \approx 0.5$ m to the cathode, the energy E collected in Δt time on an A area on the surface of the metal is $E = \eta P \Delta t \frac{A}{4\pi D^2}$. If the diameter of a metal atom is $d \approx 4$ Å and the energy required for electron emission from the given metal (the so-called work function, discussed later) is $\Delta E \approx 3$ eV, then the time needed to collect this energy by an atom is $\Delta t \approx 2.5$ s. This should have caused an observable delay between turning on the UV light and the observation of a current!

Philip von Lenard also measured the kinetic energy of the emitted electrons by applying a braking voltage U. His measurements can be simulated by an applet, which can be downloaded from http://phet.colorado.edu/en/simulation/photoelectric. At the chosen frequency and intensity of the irradiation, the voltage should be adjusted until the current drops to zero. The corresponding voltage provides the kinetic energy T in electron volts.[3] Unexpectedly, one finds that the kinetic energy does not depend on the intensity of the light at all but, as shown in Figure 3.1, above a threshold of $\nu_0 = \omega_0/2\pi$, it is proportional to the frequency ν. Instead of the energy, it is the current density that is proportional to the light intensity. These surprising results and the instantaneous emission of the electrons cannot be reconciled with classical physics.

A consistent explanation of all these findings was provided by *Albert Einstein,* assuming a continuous distribution of electron binding energies ΔE in the metal

2 A grid anode is applied to let some electrons through.

3 If $|eU| = T = mv^2/2$, the electrons cannot reach the anode anymore and no current is measured.

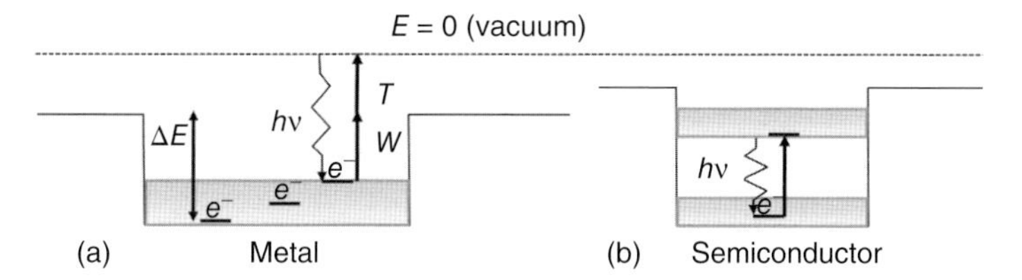

Figure 3.2 Energy scheme of the photoelectric effect in a metal (a) and of the photovoltaic effect in a semiconductor (b).

(see Figure 3.2a). The energy required to free the least bonded electron is the *work function* of the metal: $\Delta E_{\text{min}} \equiv W$. Einstein also assumed that radiation is a stream of Planck's energy quanta, $h\nu$, which can be completely and instantaneously absorbed by the electrons, provided $\nu > \nu_0 = W/h$. Then, the kinetic energy of the emitted electron becomes

$$T = h\nu - W \tag{3.1}$$

The intensity of radiation depends on the number of $h\nu$ energy quanta passing through a unit surface in unit of time. This determines the number of emitted electrons, that is, n, and so $j = env \sim I$.

This explanation has turned out to be correct and is now the basis of many practical applications.

3.2 Practical Applications of the Photoelectric Effect (Photocell, Solar Cell, Chemical Analysis)

The binding energies of the electrons are specific to the given material so the photoelectric effect can be used for the identification of the material. Since for that purpose mostly X-rays are used, this method of chemical analysis is called *X-ray photoelectron spectroscopy (XPS).*

As Figure 3.2a shows, the binding energies of the charge carriers in a metal are distributed continuously in a *band* on the energy scale. In contrast, the allowed energies of the electrons in a semiconductor are arranged in two bands (Figure 3.2b). At normal temperatures, the states of the lower band are all occupied by electrons, which are moving in all possible directions, with no resulting current. The states with energies in the upper bands are empty. After absorbing $h\nu$ energy quanta, electrons can be excited from the lower band into this empty one and – similarly to the case of a metal, where the one band is never completely filled – there can be a measurable resulting current.[4] (To make the current flow, a voltage is applied to the diode, which is a p–n junction as shown in Figure 9.1.) The generation of charge carriers in a semiconductor by light is called the *internal photoelectric effect*, or – more often – the *photovoltaic effect*. It is utilized in photodiodes (photocell), which are used not only in automatic doors but also for converting light information into information carried by

4 A more detailed explanation can only be given in the framework of semiconductor physics.

current. The photovoltaic effect is also the basis of solar cells that generate current with the help of sunlight.[5]

3.3 The Compton Effect

Einstein's successful explanation of the photoelectric effect means that the quantization of energy is not restricted to standing EM waves but is a quite general property of radiation. A little later *Arthur H. Compton* discovered that the wavelength λ of an X-ray, reflected from a metal surface, changed with the angle of observation relative to the primary beam, ϑ, as

$$(\lambda - \lambda_0) \sim (1 - \cos\vartheta) \tag{3.2}$$

where λ_0 is the wavelength of the incident radiation. In order to explain this phenomenon as well, Einstein went a step further and assumed that an energy quantum $E = h\nu$ belongs to a point-mass-like particle with momentum $p = h/\lambda$. He called these particles of light the *photons*. Then, he interpreted the reflection of the EM wave as scattering of photons by the electrons of the metal.[6]

The collision between the photon and the electron can be described in a coordinate system traveling with the original speed of the electron. In this system, the speed and momentum of the electron before the collision is zero. The momentum is always conserved in a collision, that is, $\boldsymbol{p}_0 = \boldsymbol{p} + \boldsymbol{p}_e$, where $\boldsymbol{p}_0$ and $\boldsymbol{p}$ are the momentum vectors of the photon before and after the collision, respectively, and $\boldsymbol{p}_e$ is the momentum of the electron after the collision. As shown in Figure 3.3, this corresponds to a law of cosines in the triangle defined by $\boldsymbol{p}_0$, $\boldsymbol{p}$, and $\boldsymbol{p}_e$. With Einstein's assumption about the momentum of the photon:

$$\begin{aligned} \boldsymbol{p}_e^2 = \boldsymbol{p}^2 + \boldsymbol{p}_0^2 - 2\boldsymbol{p}\boldsymbol{p}_0\cos\vartheta \Rightarrow m_e^2\mathrm{v}^2 &= \underbrace{\frac{h^2}{\lambda^2}}_{\approx\lambda_0^2} + \frac{h^2}{\lambda_0^2} - 2\underbrace{\frac{h^2}{\lambda\lambda_0}}_{\approx\lambda_0^2}\cos\vartheta \\ &= \frac{2h^2}{\lambda_0^2}(1 - \cos\vartheta) \end{aligned} \tag{3.3}$$

where it was taken into account that $\lambda - \lambda_0$ is very small.[7] In a collision of point masses, the energy is also conserved. Using $\nu = c/\lambda$ for the photon,

$$\frac{1}{2}m_e\mathrm{v}^2 = \frac{hc}{\lambda_0} - \frac{hc}{\lambda} \Rightarrow m_e^2\mathrm{v}^2 = 2hcm_e\left(\frac{1}{\lambda_0} - \frac{1}{\lambda}\right) \tag{3.4}$$

5 These devices will be described in a planned follow-up of this book, *Essential Semiconductor Physics for Engineers*.

6 N.B.: In the photoelectric effect, the photon loses all its energy, that is, it gets absorbed, while here it collides with an electron, transferring only part of its energy.

7 In a relativistic treatment (see Section A.8), this approximation is not even necessary. Before the collision, the electron has m_ec^2 energy. Afterward, its momentum is $p_e^2 = \frac{E_e^2}{2} - m_e^2c^2$. With $p_0 = \frac{E_0}{c}; p = \frac{E}{c}$ for the photon, Eq. (3.3) should be substituted by $p_e^2 = \frac{1}{c^2}[E^2 + E_0^2 - 2EE_0\cos\vartheta]$ and Eq. (3.4) by $p_e^2 = \frac{1}{c^2}[E^2 + E_0^2 - 2EE_0 + 2mc^2(E - E_0)]$. From these, with $E = \frac{hc}{\lambda}$, Eq. (3.5) follows immediately.

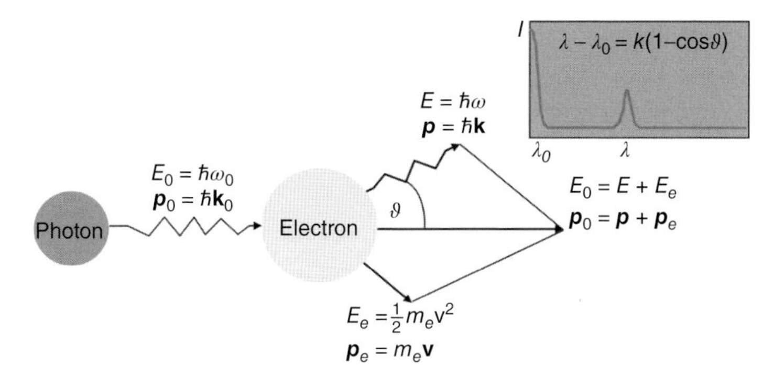

Figure 3.3 The Compton effect as photon scattering by an electron.

Comparing Eqs (3.3) and (3.4), it follows that

$$\frac{h}{cm_e\lambda_0^2}(1-\cos\vartheta) = \left(\frac{1}{\lambda_0}-\frac{1}{\lambda}\right) \approx \frac{\lambda-\lambda_0}{\lambda_0^2} \Rightarrow \lambda-\lambda_0 = \frac{h}{m_e c}(1-\cos\vartheta) \tag{3.5}$$

in agreement with the observed relation in Eq. (3.2), proving Einstein's assumptions.

3.4 The Photon Hypothesis of Einstein

The successful explanation of the photoelectric and Compton effects has justified Einstein in assuming that *light, in interaction with bodies, can be considered as a stream of point-mass-like particles,* named photons, and the relations between the wave and particle characteristics are

$$h\nu \Rightarrow E\,;\; h/\lambda \Rightarrow p \tag{3.6}$$

or, by introducing the constant *h-bar* as $\hbar \equiv \frac{h}{2\pi}$:

$$\hbar\omega \Rightarrow E\,;\; \hbar k \Rightarrow p \tag{3.7}$$

These photons behave in a collision similar to classical particles.[8] However, refraction and diffraction still cannot be explained by a stream of particles (see Section A.6). Therefore, *freely propagating light*[9] *should still be considered as a wave.* This is known as the *wave–particle duality principle,* about which a lot has been philosophized in the early twentieth century. We are going to see, however, that neither the concept of the classical point mass nor that of the classical wave can be maintained. Light does consist of photons but the properties of microscopic particles – such as the photon or the electron – are much more complicated than these simple concepts and correspond to the latter only under special conditions.

8 It should be noted that a photon has no rest mass. (That is why it can move at the speed of light.) Therefore, if it loses its total energy in an inelastic collision, as is the case in the photoelectric effect, it ceases to exist. In other words, it is absorbed by the other particle.

9 That is, light propagating without energy exchange with bodies.

3.5 Planck's Law and the Photons. Stimulated Emission

Planck's law of radiation, Eq. (2.9) describes the spectral energy density distribution of the EM field in thermal equilibrium with a body, at a temperature T, as

$$u_T(\nu) \equiv \frac{1}{V}\frac{dE_\nu(T)}{d\nu} = \frac{8\pi}{c^3h^2}\frac{(h\nu)^3}{e^{h\nu/kT}-1} \tag{3.7}$$

To prove the correctness of his assumptions, Einstein tried to derive this law by interpreting the exchange of $h\nu$ energy quanta as emission and absorption of photons. He considered three processes, as shown schematically in Figure 3.4.

- *Absorption* of a photon with energy $h\nu$ excites an atom in the wall of the metal box from its ground state E_i into an excited state $E_j = E_i + \Delta E$, where $\Delta E = h\nu$. The occurrence of such events depends on the number of atoms in the ground state, N_i, and on the number of photons with energy $h\nu$, which is given by $u_T(\nu)$. Taking B_{ij} as\ the proportionality constant, the number of absorption events is

$$N_{i\to j} = N_i B_{ij} u_T(\nu) \tag{3.8}$$

- With *spontaneous emission* of a $h\nu$ photon, atoms will return from the excited state E_j to the ground state E_i. This process depends only on the number of atoms in excited state, N_j. With the proportionality constant A_{ji}, the number of spontaneous emission events is

$$N_{j\to i}^{\text{spont}} = A_{ij} N_j \tag{3.9}$$

- With these two obvious processes, however, Einstein was not able to reproduce Planck's law. Therefore, he postulated a third one. Excited atoms should also return to the ground state if stimulated by a photon, which has exactly the energy $h\nu = \Delta E$. The energy gain of the transition from the excited state to the ground state is released by the *stimulated emission* of another photon of energy $h\nu$. The occurrence of this process depends also on N_j and on the number of $h\nu$ photons too. With the proportionality constant B_{ji} the number of stimulated emission events is

$$N_{j\to i}^{\text{stim}} = N_j B_{ji} u_T(\nu) \tag{3.10}$$

Thermal equilibrium between these three processes is achieved when

$$N_{i\to j} = N_{j\to i}^{\text{spont}} + N_{j\to i}^{\text{stim}} \tag{3.11}$$

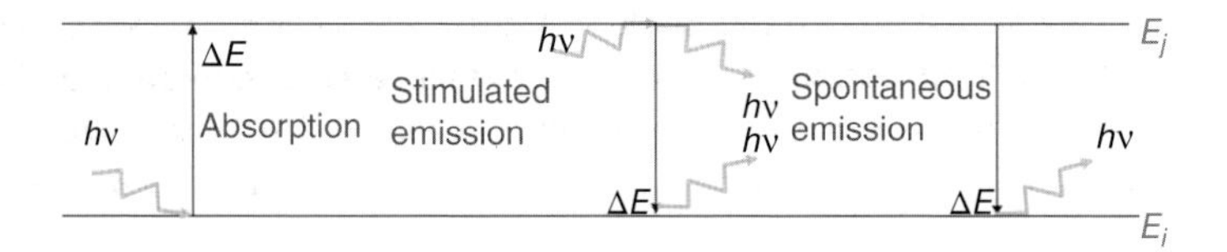

Figure 3.4 Light absorption and emission by transitions between two states of energy.

Using Eqs (3.8)–(3.10) in Eq. (3.11), it follows that

$$u_T(\nu) = \frac{\frac{A_{ji}}{B_{ji}}}{\frac{B_{ij}}{B_{ji}}\frac{N_i}{N_j} - 1} \tag{3.12}$$

The ratio N_i/N_j can be calculated by using the Maxwell–Boltzmann distribution, Eq. (A.32),

$$\frac{N_i}{N_j} = e^{\frac{E_j - E_i}{kT}} = e^{\frac{h\nu}{kT}} \tag{3.13}$$

With this, one obtains

$$u_T(\nu) = \frac{A_{ji}/B_{ji}}{\frac{B_{ij}}{B_{ji}} e^{\frac{h\nu}{kT}} - 1} \tag{3.14}$$

Comparing Eq. (3.14) with Planck's law in Eq. (3.7) shows that, for $B_{ji} = B_{ij}$ and $A_{ji}/B_{ji} = 8\pi h\nu^3/c^3$, the two equations are identical. The assumption that the EM field is a gas of light particles (photons) is in accordance with Planck's law of radiation – provided the process of stimulated emission really exists. That was then experimentally confirmed by *R. Ladenburg.* The best-known application of the stimulated emission is the laser, which is actually an abbreviation standing for: Light Amplification by Stimulated Emission of Radiation.

3.6 The Laser

Emission stimulated by an $h\nu$ photon produces a second $h\nu$ photon in the same phase as the first, so the two are coherent (capable of interference). The doubling of the number of $h\nu$ photons (Figure 3.5) means doubling the intensity. The stimulated emission leads, therefore, to a coherent amplification of the EM radiation.

This idea was first exploited for the amplification of microwave radiation (MASER), for which the American physicist *C. H. Townes* and the Russian physicists *N. G. Basov* and *A. M. Prokhorov* won the Nobel Prize in 1964. The first laser was built by *T. Maiman* in the United States in 1960.

As shown in Figure 3.6, a laser requires, first of all, an *active medium* with (at least) two well-defined energy states, E_i and E_j.

In order to increase the frequency of stimulated emission against that of the spontaneous emission, $N_j > N_i$ would be necessary. However, according to Eq. (3.13), this is not possible in equilibrium, because, in the Maxwell–Boltzmann

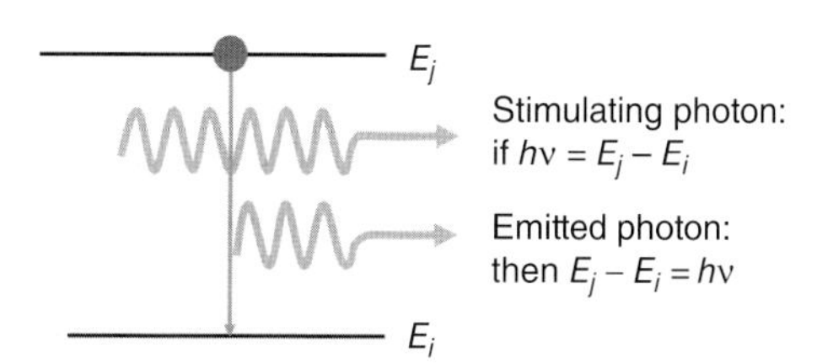

Figure 3.5 Light amplification by stimulated emission.

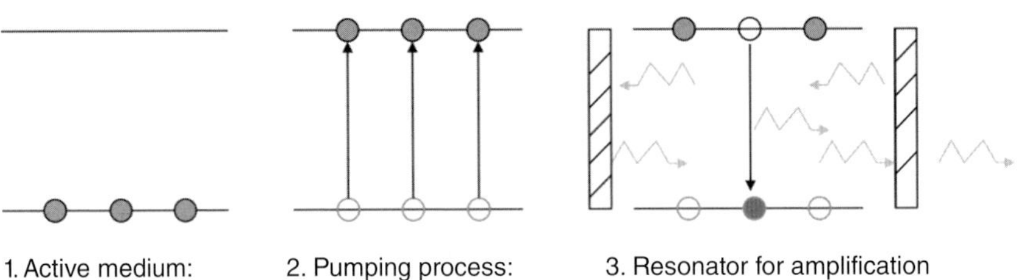

Figure 3.6 Components of a laser.

distribution, a state of lower energy is always populated by more particles than a higher energy state. So the second requirement of a laser, a *population inversion*, can only achieved by a so-called *pumping process.* Finally, to increase efficiency, each photon should stimulate many emission events. This can be achieved by a *resonator.* Essentially, this means the application of two parallel mirrors, with one of them having a reflectivity less than 100% (to allow the extraction of radiation). Keeping the photons longer in the active medium leads to a cascade of stimulated emission events, and the standing waves, arising through interference, will have a very low degree of divergence from the axial direction.

There are various ways for realizing the three components of a laser

1) *Active medium* (Figure 3.7)
   - *Two-level systems*: $N_2 > N_1$ cannot be achieved under equilibrium but currents in a *diode laser* (Figure 3.8a) make that possible.
   - *Three-level systems*: If the lifetime of the system in the state E_3 is much shorter than in the state E_2, population inversion can be achieved by $E_1 - E_3$ excitations.[10] An example of such systems is the *ruby laser* (Figure 3.8b).
   - *Four-level systems*: The efficiency can be further enhanced if the final state of the stimulated emission is emptied fast. This can be achieved by a four-level system, where the lifetimes of the states E_4 and E_2 are much shorter than that of E_3. An example for this is the *helium–neon laser* (Figure 3.8c).
2) *Pumping process*
   - *Electrical:* separate injection of charge carriers to a p–n junction of a *diode laser.*
   - *Optical:* excitation by a flash light, in *solid-state lasers* made of, for example, *ruby* or *YAG* (yttrium-aluminum-garnet).

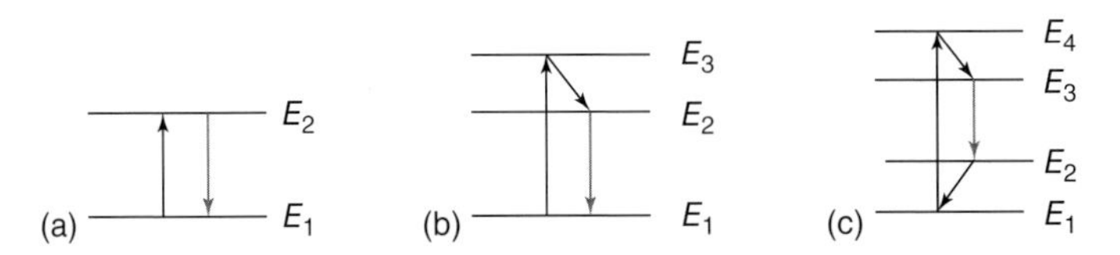

Figure 3.7 Transitions in two-level (a), three-level (b), and four-level (c) systems. The laser transition is shown in red. Color online.

10 Putting it simply, the state E_3 will be empty most of the time and accessible for excitation.

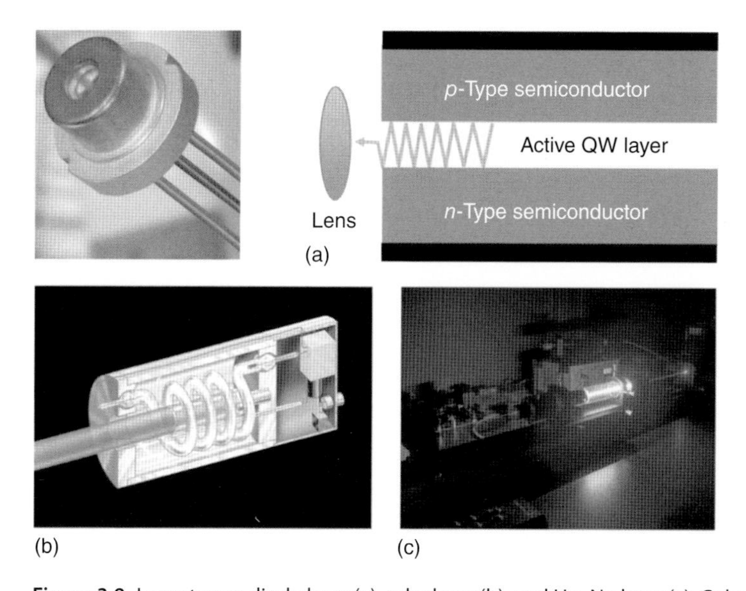

Figure 3.8 Laser types: diode laser (a), ruby laser (b), and He–Ne laser (c). Color online. (a: Reproduced with permission of Visible Diode Lasers LLC, A Florida Corporation. b: https://commons.wikimedia.org/wiki/File:Ruby_laser.jpg Used Under Creative Commons License: https://creativecommons.org/licenses/by-sa/3.0/. c: https://en.wikipedia.org/wiki/Helium%E2%80%93neon_laser Used Under Creative Commons License: CC BY-SA 3.0 https://creativecommons.org/licenses/by-sa/3.0/.)

- *Chemical:* energy exchange by the collision of different atom types, excited in a discharge tube, as, for example, in the *He–Ne* laser.

3) *Resonator*
   - *Optical mirrors* in gas and dye lasers.
   - *Polished surfaces* in diode or solid-state lasers.

Table 3.1 compares various laser types in terms of size, power, coherence length, and efficiency. The power is given both for continuous and pulsed operations. There are two types of diode lasers, with single or multiple quantum wells (the latter are called cascade lasers). The difference is explained in Chapters 9 and 10. As one can see, diode lasers are doing quite well for their sizes and can be applied in optical data storage (CD, DVD, Blu-ray disk, and also in laser printers), in optical data transfer (through a glass fiber), as well as in medicine (e.g., eye surgery) and in micromachining of semiconductor devices. For machining of metals and for remote sensing, mostly solid-state and CO_2 gas lasers are used.

Summary in Short

- In the photoelectric effect, the energy of light is absorbed in $h\nu$ quanta by electrons, while the Compton effect can be interpreted as a collision between a light particle (with energy $h\nu$ and momentum h/λ) and an electron of the metal.
- It seems that the light wave (with angular frequency ω and wave number $\boldsymbol{k}$), in interaction with bodies, behaves similar to a stream of point-mass-like

Table 3.1 Characteristics of different laser types.

Active medium	Gas	Dye	Diode		Solid state
			Single quantum well	Cascade	
Pumping	Chemical (gas discharge)	Optical	Electric (injection)	Electric (injection)	Optical
Resonator length (cm)	50	5	0.1	1	5
Power (continuous)	< 0.1 mW	< 1 W	~ mW	< 10 W	< 100 W
Power (pulsed)	1–100 kW	100 kW	10 W	—	Up to GW
Coherence length	High	Low	Low	High	Low
Efficiency (%)	0.1–20	0.5–15	1–20	1–20	0.01–5

particles of energy $E = \hbar\omega$ and momentum $\boldsymbol{p} = \hbar\mathbf{k}$. These particles are called *photons*. Planck's law of blackbody radiation can be derived with the help of photons assuming three processes:

1. Bodies can instantaneously and completely absorb photons, achieving an excited state of higher energy (*absorption*).
2. Bodies in the excited state can return to the ground state spontaneously by emitting a photon (*spontaneous emission*).
3. Besides being absorbed, photons can also induce the body to return to the ground state by emission of another photon (*Stimulated emission*), achieving light amplification.

- The relative probabilities among the aforementioned three processes prefer the population of the ground state in equilibrium but, by using various pumping processes, a population inversion can be realized. Applying a resonator system, a cascade of stimulated emission events can be achieved, leading to strong radiation with little divergence from the optical axis. This coherent amplification is the *laser.*

3.7 Questions and Exercises

Problem 3.1 Consider the interaction of light (traveling in vacuum) and a material with two well-defined energy states (two-level system). The photon energy is 3 eV.

a) What is the light frequency? Calculate the wavelength from the dispersion relation. What is the momentum of the photons?
b) If the absorption constant is $B_{ij} = 10^9(\mathrm{m \cdot kg^{-1}})$, what are the constants of the spontaneous and of the stimulated emission, A_{ji} and B_{ji}, respectively? (Hint: compare Eqs (3.7) and (3.4)!)

c) Demonstrate that population inversion is not possible in thermal equilibrium of a two-level system. (Hint: Use the Maxwell–Boltzmann distribution of Section A.7!).

Problem 3.2 An X-ray with the wavelength of $5 \cdot 10^{-12}$ m is scattered by electrons. The recoiling electron has the energy of 70 keV. Under what angle will the photon be reflected and what will be its wavelength?

Problem 3.3 What is meant by the wave–particle duality of light? What experiments have made this assumption necessary?

4

Electrons: The Physics of the Discharge Lamps

In this chapter...
As we saw in the previous two chapters, studying the physics of the light bulb has eventually led to the discovery of laser. In addition to the practical consequences, however, this research has also contributed to the development of physics itself. The way how classical physics has considered light was questioned. On the one hand, light interference could still only be explained by the assumption of infinite harmonic waves propagating in the continuum of the electromagnetic (EM) field. On the other hand, in interaction with bodies, light seemed to be a stream of point-mass-like particles (photons). In this chapter, we see how the detailed investigation of discharge lamps can lead to the claim that the smallest particles of a body, considered as point masses in classical physics, may sometimes behave like waves.

4.1 Fluorescent Lamp

A common fluorescent lamp (CFL) is a discharge tube, with an electric switch to ignite the discharge in the gas fill (see Figure 4.1). The electric switch is a very important component, which needs to be optimized to minimize energy losses and the time necessary until achieving full light intensity[1]; however, here we concentrate only on the discharge tube.

The operational principles of a fluorescent lamp have already been described briefly in Chapter 1. The voltage between the electrodes extracts electrons from the heated cathode. The electrons collide with the atoms of the gas fill and ionize them; this is the plasma discharge. A cascade of electrons reaches the anode and the heat, produced by the current, evaporates the mercury droplet in the tube. The mercury atoms are then excited by colliding electrons. Returning to the ground state, they emit UV light, which is absorbed by the "phosphor" sheet on the inner wall of the tube. The absorbed energy excites the molecules of the phosphor and, upon returning to the ground state, they emit whitish light (fluorescence).

---

1 It should be noted that the switching time of the discharge lamps is still significantly longer than that of the light bulb.

Essential Quantum Mechanics for Electrical Engineers, First Edition. Peter Deák.

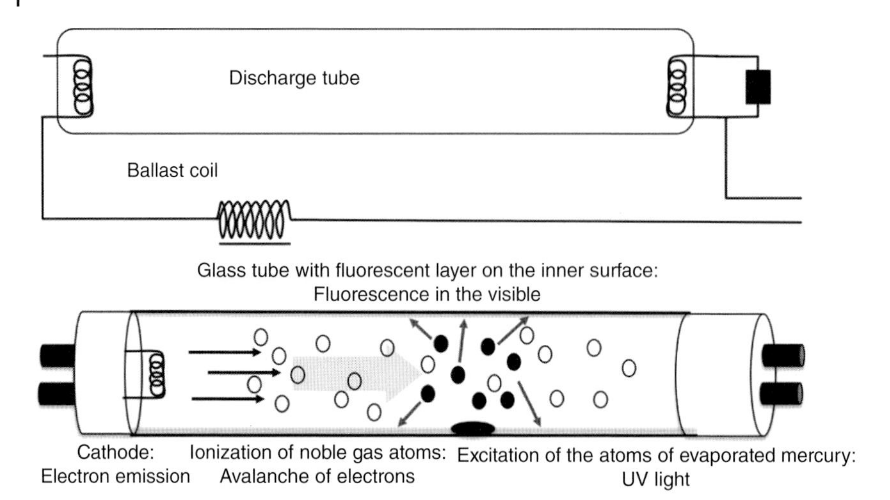

Figure 4.1 Structure of a discharge lamp.

Without mercury, the gas fill (mostly a krypton–xenon mixture) emits only green light. A neon fill would emit red light. Chemists know very well that excited atoms of each element emit a characteristic frequency. For example, the color of a flame becomes yellow when strewing sodium in it. The question arises: how the structure of atoms can explain the specific color?

4.2 Franck–Hertz Experiment

A clue to the answer to the aforementioned question has been provided by the investigation of the I–V characteristics of a DC-driven discharge tube with neon or mercury fill. The experiment, originally carried out by *J. Franck* and *G. Hertz*,[2] can be simulated by the applet at http://phys.educ.ksu.edu/vqm/free/FranckHertz.html. The experimental setup and the results are displayed in Figure 4.2.

Above a critical voltage between the heated filament cathode and the grid, $V_G \geq V_0$ (in the applet ≈ 1 V), electrons are emitted from the cathode and a current starts flowing. With increasing V_G, this current increases linearly according to Ohm's law, but only up to another critical voltage $V_1(\approx V_0 + 1.8$ V for the neon fill and $\approx V_0 + 4.4$ V for mercury). Above that, the current drops suddenly and, at the same time, gas atoms near the grid start to emit light: red by neon and UV by mercury. The frequency is $\nu \approx (V_1 - V_0)/h$. Increasing the voltage further leads again to a linear increase of the current and the light emission of the atoms shifts toward the cathode. After reaching $V_2 \approx (2V_1 - V_0)$, the current drops again and one observes two emitting regions: one halfway between the cathode and the grid and another near the latter. The phenomenon is repeated for $V_3 \approx (3V_1 - V_0)$, and so on, with ever more lighting regions.

2 The nephew of H. Hertz.

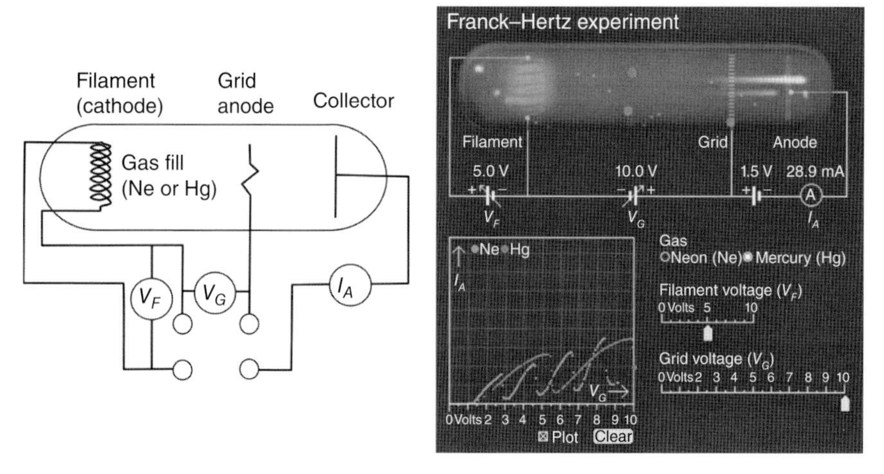

Figure 4.2 The Franck–Hertz experiment. Color online. (Snapshot from the applet of the Kansas State University Physics Education Research Group, http://phys.educ.ksu.edu/vqm/free/FranckHertz.html, with the kind permission of Prof. Dean Zollman.)

These observations can be explained in the following way: as we have already seen at the photoelectric effect, the energy, corresponding to the work function of the electrons, $W = qV_0$, has to be invested (here $q = -e$). The emitted electrons are exposed to a field strength of $E_x = (V_G - V_0)/l$ between the cathode and the grid, with l being the distance between them. Over the path x, the work of the force qE_x provides the electrons with the kinetic energy

$$T = \frac{1}{2}m_e v^2 = qE_x x = q(V - V_0)\frac{x}{l} \tag{4.1}$$

As long as the kinetic energy of the electrons remains below the value of $q(V_1 - V_0)$ up to $x = l$, the atoms cannot absorb energy and the collisions between the electrons and the atoms remain elastic. If the voltage is high enough for the electrons to reach this critical energy at $x \approx l$, the collisions become completely inelastic. The electrons lose all their kinetic energy and cannot reach the grid. So the current drops suddenly. The atoms near the grid, which are hit by the electrons, are excited by the absorbed energy. This leads then to the spontaneous emission of photons with the energy $h\nu = q(V_1 - V_0)$. For a higher voltage, the critical energy is obtained by the electrons already at $x < l$, the emission shifts nearer to the cathode and, after the collision, the electrons can again collect energy over the path $(l - x)$ and reach the grid. The current rises linearly again. For $q(V_2 - V_0)/2 = h\nu$, the kinetic energy of the electron becomes sufficient for an adsorption/emission event halfway between cathode and the grid, and the energy collected afterward is sufficient for another one near the grid. To put it briefly, the atoms can only absorb a specific amount of energy, the quantum $q(V_1 - V_0)$, and then emit a photon with the frequency $\nu = q(V_1 - V_0)/h$. This frequency is characteristic to the atom and that explains the monochromatic radiation of a discharge lamp, filled with one type of atoms only. It should be noted that, actually, each atom can absorb at several discrete frequencies but only a very few of those are in the visible region of the spectrum.

The Franck–Hertz experiment confirms the assumption of Planck, that is, bodies (in this case, atoms of a gas) can lose energy to the EM field only in quanta. Now the question arises: Why cannot the atoms absorb just any energy in the collision with the accelerated electrons of the discharge tube? What happens with the absorbed energy? To answer these questions, one should have a model of the atom, that is, an idea about its structure.

4.3 Bohr's Model of the Hydrogen Atom: Energy Quantization

The simplest atom is hydrogen, which also emits only discrete frequencies. As shown in Figure 4.3, the emission from a discharge tube, filled with hydrogen, can be analyzed by a diffraction grating (see Section A.6.2). The emission spectrum contains a few lines in the visible and further ones in UV and IR. Interestingly, the observed frequencies can be calculated from the formula found by *J. J. Balmer:*

$$\nu = R' \left(\frac{1}{m^2} - \frac{1}{n^2} \right) \tag{4.2}$$

where $m < n$ are integers and R' is a constant. Considering that the frequency is connected to the emitted energy as $\nu = \Delta E/h$, this means that the atom can only have states with discrete energies, $E_n \sim 1/n^2$.

For a long time, atoms have been thought of as undividable building blocks of the chemical elements. (The Greek word "*atomos*" actually means *undividable.*) However, *E. Rutherford* observed that the so-called α-rays (a stream of positively charged helium nuclei) can penetrate an aluminum foil, and only a few of the ions will have trajectories deviating from a straight line. Considering those trajectories, *Rutherford's atom model* was proposed: an atom consists of a tiny, positively charged nucleus and of even tinier electrons moving about in the electric field of the nuclei. (This means that most of the volume of an atom is "empty," and the ions will only be scattered if they fly close by a nucleus. The electrons, which are much lighter, can hardly scatter them.) Since the Coulomb force (see Eq. (A.10)) is, similarly to gravitation, proportional to $1/r^2$, the orbits of the electrons should be ellipses or circles, analogously to the planetary motion around the sun. According

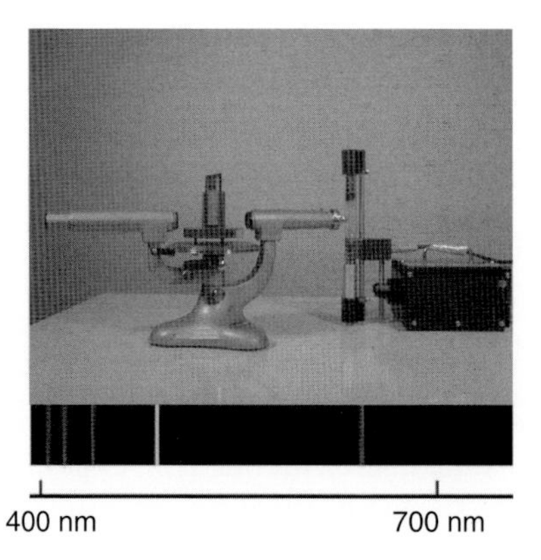

Figure 4.3 Measuring the emission of a hydrogen plasma. The spectrum is shown at the bottom. Color online. (Reproduced with permission of G. Brackenhofer, from The Lecture Collection of the University of Ulm.)

to the Newtonian equation of motion for a circular orbit, the centripetal acceleration (see Eq. (A.4a)) of the electron with mass m_e is caused by the Coulomb force between the electron (charge $-e$) and the proton (charge $+e$) in the nucleus of a hydrogen atom:

$$-m_e\frac{v^2}{r} = -\frac{1}{4\pi\varepsilon_0}\frac{e^2}{r^2} \tag{4.3}$$

at *any radius r*. With the potential energy of the Coulomb field (Eq. (A.11)), the total energy can be written as

$$E = \frac{1}{2}m_e v^2 - \frac{1}{4\pi\varepsilon_0}\frac{e^2}{r} = \frac{1}{2}\frac{1}{4\pi\varepsilon_0}\frac{e^2}{r} - \frac{1}{4\pi\varepsilon_0}\frac{e^2}{r} = -\frac{1}{2}\frac{1}{4\pi\varepsilon_0}\frac{e^2}{r} \tag{4.4}$$

where Eq. (4.3) has been used to express $m_e v^2$. Since there is no restriction on r, there cannot be one on the energy either: it can change continuously. This, however, contradicts the experience with the discharge lamp.

The miniature solar system of the Rutherford model might appear very cute, but it is physically unacceptable anyhow. An electron on circular orbit has a centripetal acceleration and, as known from electrodynamics and also from the practical example of an antenna, an accelerated charge is radiating. This means a loss of energy and, according to Eq. (4.4), an ever smaller radius until eventually the electron falls into the nucleus. So, a classical model fails not only to explain the quantization of the energy but it does not provide a working description for the motion of the electron in the atom either.

The next step of development was the atom model of *Niels Bohr*. Bohr's atom model is based on two assumptions that *do not follow from classical physics:*

a) Electron orbitals in an atom, with quantized angular momenta

$$L = |r \times m_e v| = r m_e v = n\hbar; \quad n = 1,2,\ldots, \tag{4.5}$$

are stable, that is, the electron does not radiate (despite the centripetal acceleration).

b) During transition from one such orbital n, to another m, the electron emits or adsorbs the energy difference (see Figure 4.4):

$$h\nu = E_m - E_n \tag{4.6}$$

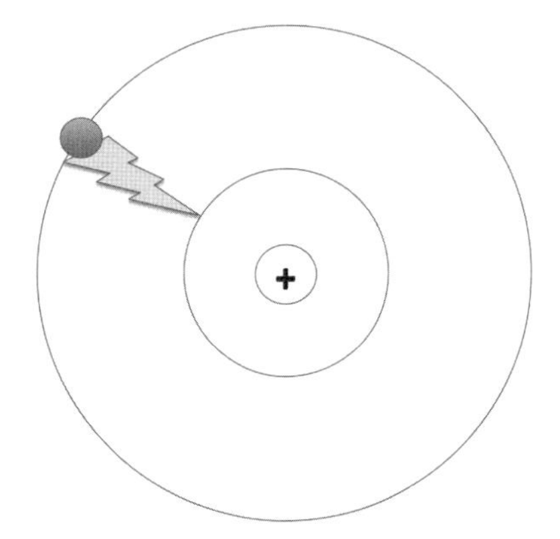

Figure 4.4 Bohr's model of the hydrogen atom.

Taking Eq. (4.5) into account in Eq. (4.3) leads to a quantization of velocity,[3]

$$v_n = \frac{e^2}{4\pi\varepsilon_0 n\hbar} \tag{4.7}$$

and to discrete (quantized) radii,[4]

$$r_n = \frac{4\pi\varepsilon_0\hbar^2}{m_e e^2} n^2 \tag{4.8}$$

Using Eqs (4.7) and (4.8) in Eq. (4.4) gives then a quantized energy

$$E_n = -\frac{me^4}{8\varepsilon_0^2 h^2}\frac{1}{n^2} = \frac{\overbrace{-13.605}^{R}}{n^2}(\text{eV}) \tag{4.9}$$

where the Rydberg energy, $R = 13.605$ eV, has been introduced. Using this expression in Eq. (4.6) gives

$$\nu = \frac{R}{h}\left(\frac{1}{m^2} - \frac{1}{n^2}\right) \tag{4.10}$$

This result explains the Balmer formula and, according to measurements, the constant R' is indeed equal to R/h. So whatever the reason for the quantization of the angular momentum (and for the lack of radiation from such an orbital), the Bohr model is in quantitative agreement with the observed spectrum of the hydrogen atom. Note that we have considered the motion of an electron in the field of a single proton in Eq. (4.3). If the number of protons in the nucleus was Z, that is, its charge was $+Ze$, then our derivation would lead to Z^2e^4 in Eq. (4.9) instead of e^4, and to Z^2R in Eq. (4.10) instead of R. This explains why the allowed frequencies change from atom to atom with the atomic number Z. The smallest energy that can be emitted by mercury is larger than that of neon.[5]

4.4 Practical Consequences of the Energy Quantization for Discharge Lamps

While Bohr's quantization of the angular momentum cannot be explained within classical physics, the quantization of the electron energy in the atoms following from it is obviously a fact. One has to accept that the choice of the gas fill determines the color of a discharge. This is the reason why the discharge lamps filled with noble gases or the low-pressure sodium lamp have a single color and cannot be used for interior lighting. The low-pressure sodium lamp has a deep yellow

---

3 After multiplication by r^2 it follows from Eq. (4.3) that $(r \cdot m_e v)\cdot v = \frac{e^2}{4\pi\varepsilon_0}$

4 Substituting v_n from Eq. (4.7) into Eq. (4.5) gives Eq. (4.8).

5 Unfortunately, the Bohr model is quantitatively correct for the hydrogen atom only and cannot be extended to other atoms. Even in the case of hydrogen, it gives only the energies right but not the angular momentum and the magnetic momentum in the ground state (see Chapter 12). The Bohr model is a step in the right direction, but it is based on the classical concept of the point mass and is, therefore, essentially wrong.

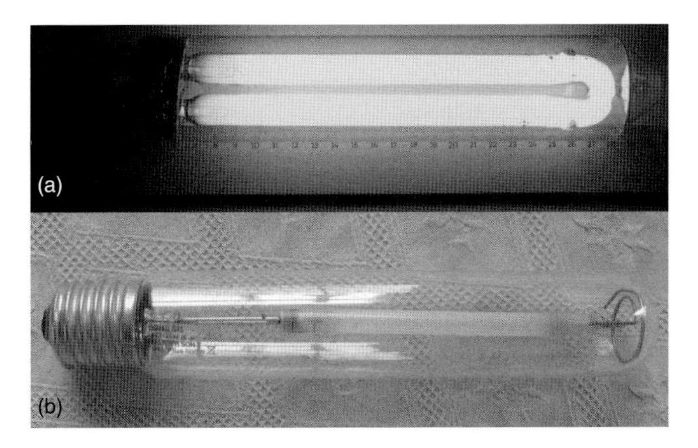

Figure 4.5 Low- (a) and high- (b) pressure sodium lamps. Color online. (https://en.wikipedia.org/wiki/Sodium-vapor_lamp Used Under Creative Commons License: CC BY-SA 3.0 https://creativecommons.org/licenses/by-sa/3.0/.)

color (Figure 4.5a). In high-pressure sodium lamps (Figure 4.5b), a very high temperature is reached in an Al_2O_3 discharge tube, and the thermal broadening of the emission line of sodium makes the color a bit more similar to that of the sun. Such lamps are used for exterior lighting (streets and stadiums) but are not suited for lighting of living quarters.

High-pressure metal-halide lamps have a quartz discharge tube, allowing for high temperature operation, and contain a mixture of various molecules. While simple molecules also emit discrete frequencies only, with a mixture of them at high temperature (with broadened emission lines), one can get more or less white light, alas with a rather rugged spectrum (see Figure 4.6). This might be advantageous for plants but not for the human eye.

Actually, the only acceptable alternative for interior lighting is fluorescent lamps (Figure 4.1), which are, however, quite environmentally unfriendly because of the use of mercury. The color conversion (from UV to white) is based on the many allowed energies of electrons in the complicated molecules of the phosphor. As shown in Figure 4.7, the conversion proceeds in three steps:

Step 1: *Absorption* of a UV photon and excitation of an electron.
Step 2: *Thermalization*, that is, energy loss of the electrons by exciting molecular vibrations.
Step 3: *Luminescence*, that is, emission in the visible after a delay of Δt time. Depending on the lifetime of the excited state, this is called
  (a) fluorescence, if $\Delta t < 10^{-8}$ s,
  (b) phosphorescence, if $\Delta t < 10^{-8}$ s.

Fluorescent "phosphors" are applied in CFLs and in displays, while phosphorescence is utilized in some LEDs.

Actually, the spectrum of fluorescence lamps is not very smooth either. Figure 4.8 compares it with the spectrum of the sun, measured at sea level.

In fact, lighting with discharge lamps are for the time being, without exception, suboptimal for our color recognition. They are, however, economically more

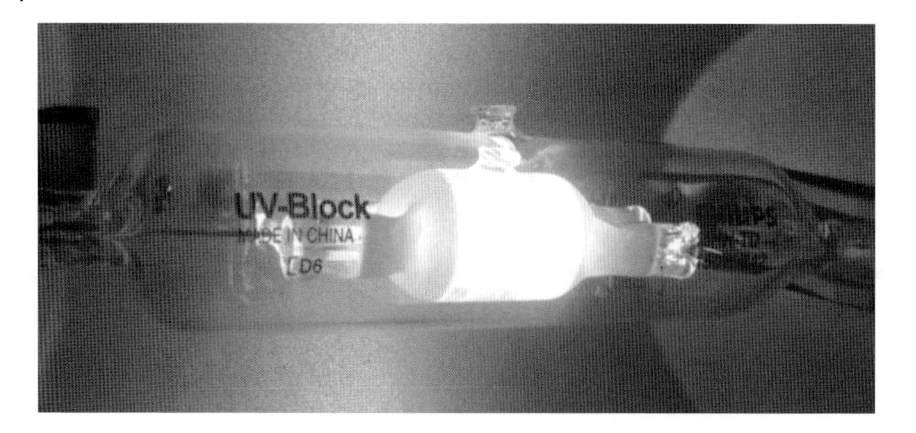

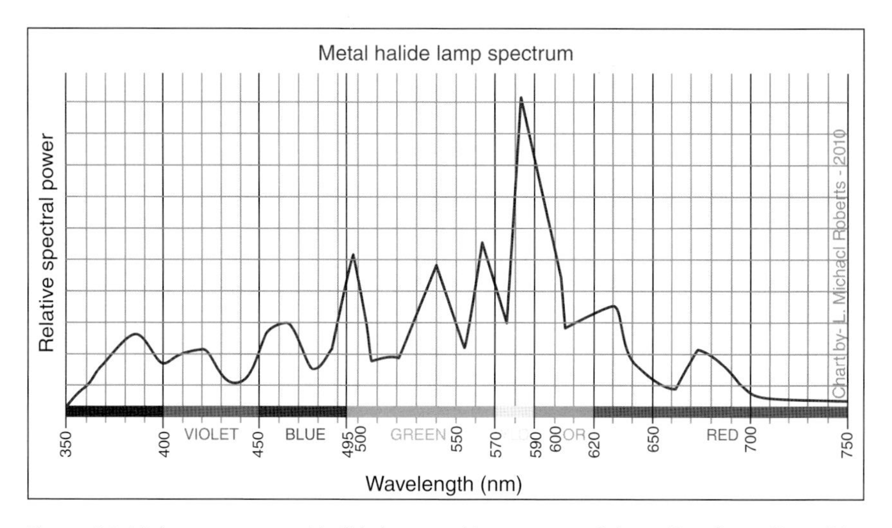

Figure 4.6 High-pressure metal-halide lamp and its spectrum Color online. https://en.wikipedia.org/wiki/Metal-halide_lamp Used Under Creative Commons License: CC BY-SA 3.0 https://creativecommons.org/licenses/by-sa/3.0/. The spectrum was taken by L. Michael Roberts.)

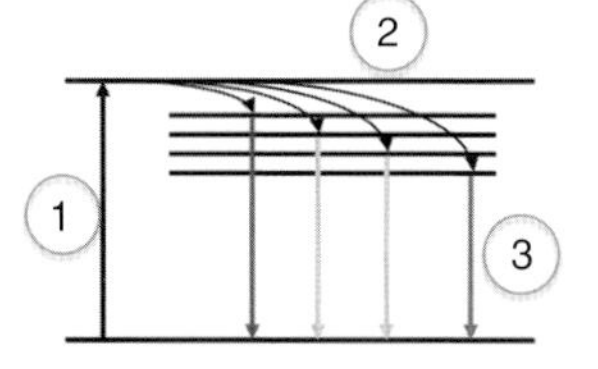

Figure 4.7 Scheme of the light emission by phosphor. Color online.

viable than light bulbs. Even though the energy loss at igniting the plasma discharge can be as high as 20–30%, and about the same amount is lost in the excitation and color conversion, discharge lamps still have a much better lumen/watt ratio than incandescent lamps, as can be seen by comparing the values in Table 4.1 to those in Table 3.1. From the development between 2000 and 2010, one can guess that a substantial improvement of discharge lamps cannot be expected anymore either.

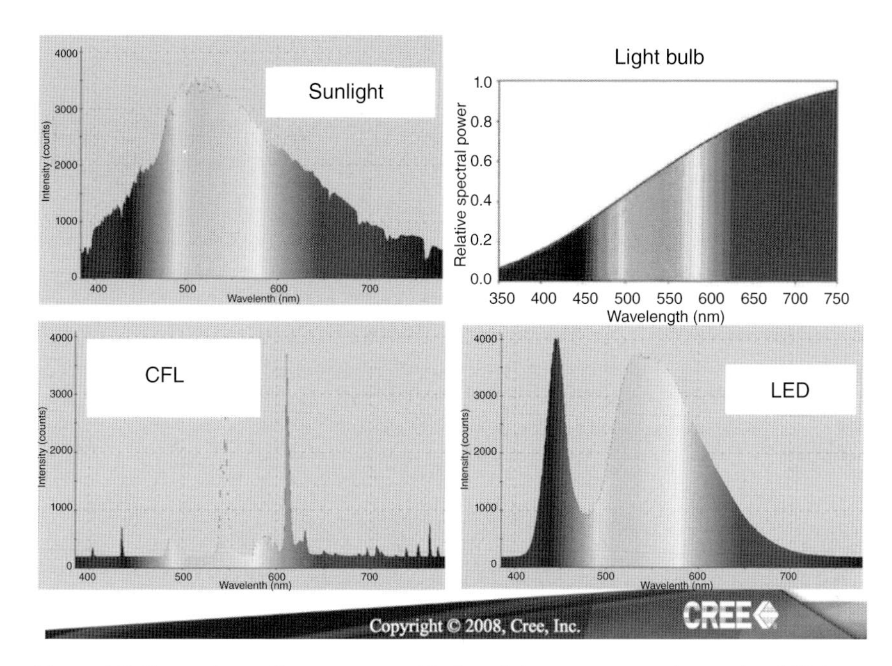

Figure 4.8 Comparison of lamp spectra with that of the sun measured at sea level. (Color online.) (Reproduced with permission of Mitch Sayers, Cree Inc.)

Table 4.1 Characteristics of discharge lamps for interior lighting.

	Power consumption (W)	Electric loss (%)	Efficiency 2000 ($lm W^{-1}$)	Efficiency 2010 ($lm W^{-1}$)	Lifetime (h)
CFL	10–60	20–30	65–100	70–100	10 000
Metal-halide	35–70	20–30	45–65	50–65	8 000

This leaves us with LEDs, which have a substantially smoother spectrum (Figure 4.8). The LED was born out of our understanding of the physics of electrons in a solid, that is, out of the quantum mechanics. Therefore, we should not be content with the unexplained assumptions of the Bohr model. As we discuss later, the search for the reason for the assumptions in the Bohr model has led to a deeper understanding of the world and to the discovery of micro- and optoelectronics.

4.5 The de Broglie Hypothesis

In the atom model of Bohr, quantization of energy follows from quantization of angular momentum. But why should the latter be quantized in the first place? After all, angular momentum is defined by the position and the velocity vector, both of which are continuous quantities. In an attempt to find an explanation, *Louis de Broglie* assigned wave-like behavior to the electron, using Einstein's

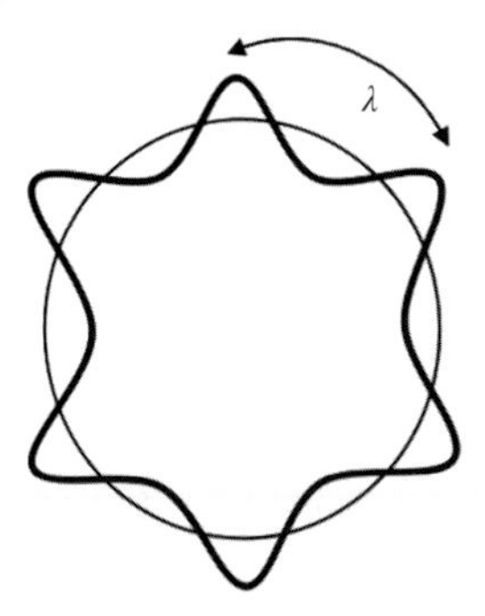

Figure 4.9 Stationary wave along a circle.

assignment in Eq. (4.9) the other way round:

$$E \Rightarrow \hbar\omega\ ;\ p \Rightarrow \hbar k \tag{4.11}$$

If such an electron wave is forced on a circular orbit of radius r, it will interfere with itself. The wavelengths of the resulting stationery waves must satisfy the condition

$$2\pi r = n\lambda;\quad n = 1,2, \ldots \tag{4.12}$$

(see Figure 4.9).[6]

Since $\lambda = 2\pi/k$, from Eq. (4.12) it follows that $kr = n$. After multiplication with $\hbar$, and considering Eq. (4.11), that is, $p = \hbar k$, the definition of angular momentum of a circular orbit $|L| = rp$ leads to

$$L = \hbar n;\quad n = 1,2, \ldots \tag{4.13}$$

which is exactly the assumption (a) of the Bohr model, that is, Eq. (4.5). Therefore, *if we accept that the electron with energy E and momentum p behaves like a wave with $\omega = E/\hbar$ and $k = p/\hbar$, the quantization of the angular momentum follows immediately.*

The electron is an elementary particle and should be the nearest thing to a point mass. The whole classical physics is based on this assumption! Should the mental image of an electron as a tiny sphere be abolished now? Before doing so, we should seek direct experimental confirmation for the wave assumption of de Broglie.

4.6 The Davisson–Germer Experiment

If the electron is a wave, it should produce an interference pattern behind a diffraction grating. A constant accelerating voltage U can define the momentum of the electron (see Eq. (A.8)) by

$$eU = T = p^2/(2m_e) \tag{4.14}$$

From that, using Eq. (4.11), follows the *de Broglie wavelength* of the electrons:

$$\lambda = \frac{h}{\sqrt{m_e eU}} \approx \frac{1.23\,(\text{nm}\cdot\text{V}^{1/2})}{\sqrt{U}} \tag{4.15}$$

6 If Eq. (4.12) is satisfied, the wave will reach the same point after one circle in phase, so there will be no extinction.

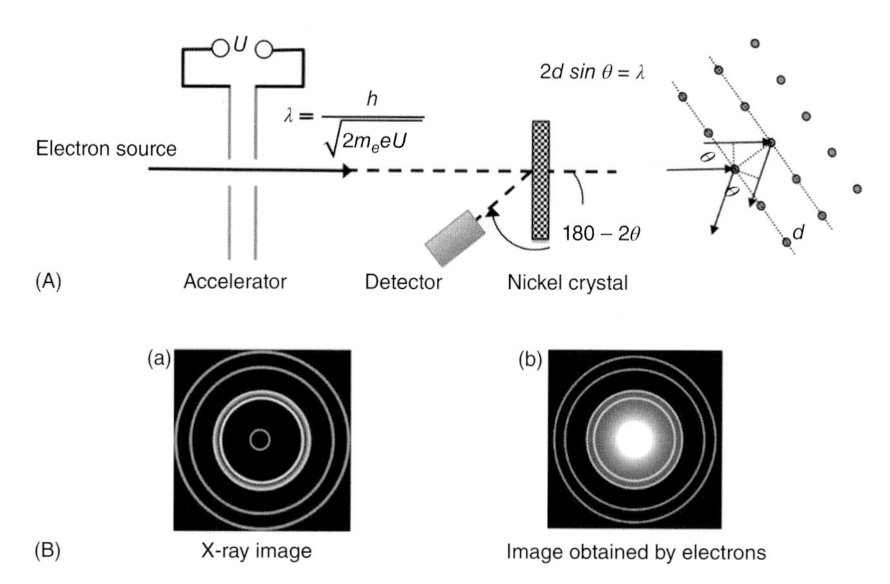

Figure 4.10 (A) Schematic depiction of the Davisson–Germer diffraction experiment with electrons. (B) Diffraction pattern obtained by an X-ray (a) and by electrons (b).

With $U \approx 100$ V, the wavelength becomes comparable to the atomic distances in a crystal. Correspondingly, one should get the same diffraction pattern with electrons as with X-rays. A detector should register electrons at the *Bragg* angle θ, satisfying the condition (see Section A.6.2)

$$2d \cdot \sin\theta = n\lambda \tag{4.16}$$

where d is the distance of the atomic planes in the crystal. Such an experiment was carried out by *C. Davisson* and *L. Germer* on a polycrystalline nickel sample and, as shown in Figure 4.10, the diffraction image was identical to the one obtained by applying an X-ray.

4.7 Wave–Particle Dualism of the Electron

The Davisson–Germer experiment is an unambiguous confirmation of the wave nature of the electron. However, if the electron was a material wave extending in space, it could be divided (similar to a water wave by inserting a wall into the middle of a ripple tank). This runs against the Millikan experiment, which clearly excludes the existence of a charge smaller than that of the electron, that is, the electron must be undividable. This controversy is similar to the case of light and was also resolved (temporarily) by the wave–particle duality principle. *It seems as if both the elementary excitations of the EM field and the elementary particles of a body can behave under specific circumstances as point masses and under other circumstances as waves.* In the following chapter, we see that actually none of these two classical concepts can be sustained.

Summary in Short

- The Franck–Hertz experiment (measurement of the electron current as a function of the anode voltage in a discharge tube filled with one type of gas) has shown that atoms can only absorb energy in specific quanta, and they emit the absorbed energy as photons.
- The frequencies emitted by atomic hydrogen follow the Balmer formula, $\nu \sim (1/m^2 - 1/n^2)$, where $m < n$ are integers.
- The atom model of Rutherford, a miniature solar system with electrons orbiting the nucleus under the effect of the Coulomb force, is unacceptable. Since an accelerated charge radiates, the centripetal acceleration would make the electron lose energy and it would not be able to maintain a constant radius around the nucleus.
- In Bohr's model, it is assumed that an electron on orbitals with angular momenta $L = n\hbar$ $(n = 1,2, \dots)$ does not radiate for some reason. This assumption leads to a quantized energy of the orbitals:

$$E_n = -\frac{me^4}{8\varepsilon_0^2 h^2}\frac{1}{n^2} = \frac{-13.605 \quad \text{(eV)}}{n^2}$$

- Bohr also assumed that in an electronic transition from orbital m to n, the energy difference is emitted as a $h\nu$ photon. The assumptions of Bohr can explain Balmer's formula, in quantitative agreement with experiment.
- According to the de Broglie hypothesis, the quantization of the angular momentum follows from the self-interference of the electron wave (with $\omega = E/\hbar$ and $k = p/\hbar$) on a circular orbit.
- The Davisson–Germer experiment proves that electrons are indeed capable of interference. The same diffraction pattern can be obtained from a crystal by an X-ray or an electron beam. This proves the wave nature of the electron.
- However, electrons are undividable elementary particles. Obviously, the question of the wave–particle duality arises just as it did for light.
- From the quantization of energy in atoms, it follows that any elementary gas in a discharge lamp can emit only a few discrete frequencies in the visible. To obtain white light, many different molecules must be present in the fill (such as in the metal-halide lamp) or in the phosphor of fluorescent lamps. The spectrum, however, will be quite rugged, which impairs our color recognition. Fluorescent lamps are from this respect passable but contain mercury.

4.8 Questions and Exercises

Problem 4.1 According to Bohr's model of the hydrogen atom, what wavelength should a photon have to

(a) excite an electron from the ground state into the third allowed energy level?
(b) to ionize the atom when the electron is already in this excited state?

Problem 4.2 Apply Bohr's formulas to the He^+ ion.

(a) Write down the equations for the radius and energy of the orbitals and calculate the second smallest radius!

(b) What is the relation between the frequencies of the hydrogen spectrum and of the spectrum of He^+?

Problem 4.3 Consider the helium atom in the framework of the Bohr model.

(a) What would be the energy of a He atom in the ground state (two electrons in orbital $n = 1$), if the Coulomb repulsion between the electrons could be neglected?
(b) Estimate the effect of the Coulomb repulsion between the electrons, $V_C = e^2/(4\pi\varepsilon_0 r_{12})$, assuming that their distance r_{12} is the longest possible on the $n = 1$ orbital. What would be the total energy of the ground state with this estimate? (N.B.: Comparison with the experimental value of −78.9 eV shows the error of the Bohr model.)

Problem 4.4 The quantization of the angular momentum can be derived from the assumption that the electron wave must be stationary on a circular orbit. Use this assumption and the Bohr radii of hydrogen to calculate the momentum and frequency of the electron on the first two orbitals ($n = 1$ or 2).

Problem 4.5 An electron beam, accelerated by the voltage $U = 63{,}64$ V, is diffracted by a copper crystal. A diffraction maximum ($n = 1$) is observed at the angle of 25°. What is the distance between the reflecting atomic planes?

Problem 4.6 What does the wave–particle duality principle say and what experiments have made its formulation necessary?

5

The Particle Concept of Quantum Mechanics

In this chapter…
We are going to scrutinize the idealized models of classical physics for bodies and radiation, namely point masses and waves, respectively. As we have seen in Chapters 3 and 4, light and elementary particles can show both point-mass-like and wave properties. Therefore, we consider a decisive experiment in which the ability of a *single electron* for interference is investigated. The interpretation of the experimental results will lead to the conclusion that subatomic particles can have many more than just two forms of existence. We cannot even imagine these states, but none of them really corresponds to the classical concepts of point masses and waves. The way out is the introduction of a mathematical function, called the *wave function*, which provides a complete description of the state of a subatomic particle and allows the prediction of its properties. As we discuss later, this wave function can be determined with knowledge of the potential energy function.

5.1 Particles and Waves in Classical Physics

Classical physics can essentially be divided into two parts according to the two sides of the Newtonian equation of motion (Figure 5.1). If the force is known, this equation allows the prediction of the position and the velocity of a point mass, as a function of time (assuming, of course, that the initial conditions are known). This is the task of mechanics.[1] The force describes the interaction of the point mass with another, via a force field. The laws of basic force fields are formulated in the field theories (gravitation and general relativity theory, Maxwell's theory of the EM field, nuclear and particle physics). If we do not consider celestial bodies (for which gravitation is important) or particles in the nucleus of an atom (bound by strong and weak interactions), the decisive force field is the electromagnetic one, described by Maxwell's axioms. (Most known forces, such as the ones responsible for elasticity, friction, or drag, are consequences of electromagnetic interactions.)

1 Mechanics includes also the theory of special relativity. Statistical physics (which lays the foundation for thermodynamics) describes the distribution of positions and velocities in a system of many point masses.

Essential Quantum Mechanics for Electrical Engineers, First Edition. Peter Deák.

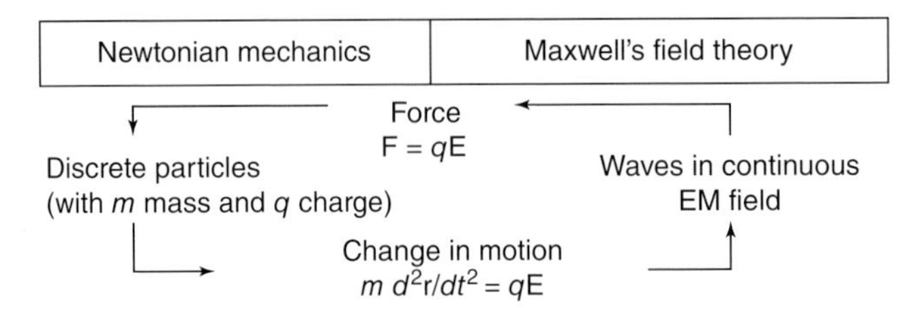

Figure 5.1 The two main parts of classical physics.

According to the Newtonian equation of motion, a point mass m with charge q is accelerated by the electric field **E**. According to Maxwell's theory, however, an accelerated charge radiates electromagnetic (EM) waves, which will change the electric field.[2]

Newton's equation considers discrete particles (bodies) and idealizes them as point masses. The corpuscular structure of bodies and the idealization of individual particles as point masses are supported by many experiments (from Galilei to Millikan, see Chapter 1). Maxwell's theory assumes a continuous EM field in which waves (radiation) can propagate. The wave nature of light (postulated by Huygens and Fresnel, see Chapter 1) is confirmed by countless interference experiments. Even though Einstein's theory of special relativity has shown that a body (mass) can be transformed into EM radiation and vice versa (see Eq. (A.46)), the basic concepts, applied in classical physics for bodies and radiation, are quite different (see Figure 5.2).

Point masses can transfer energy by convection and conduction,[3] while energy is transferred in the EM field by radiation. However, the main difference between the concept of a *localized point mass* and the concept of a *delocalized wave* can be best seen in a diagram, displaying the probability to find them at any point in space. (Figure 5.2 shows such a diagram in one dimension (1D), along the x-axis.) In the case of a point mass with the trajectory $x(t)$ and momentum $p(t)$ – both a continuous function of time – the $W(x)$ probability to find it at x is 1 at a given point in space (where the point mass actually is), and 0 everywhere

Body: point mass	Radiation: wave
$E = mv^2/2$	$E = \varepsilon_0 c E^2 V \Delta t$
Probability distribution $W(x)$ … $x(t), p(t)$ … x	Probability distribution $W(x)$ … ω, k … x
Experimental confirmation: Galilei & Millikan	Huygens & Fresnel

Figure 5.2 Concepts of classical physics.

2 We have not considered here the effects between a moving charge and the magnetic field.
3 The kinetic energy of a point mass can be transferred in space by the motion of the particle, or through collisions with other particles, respectively.

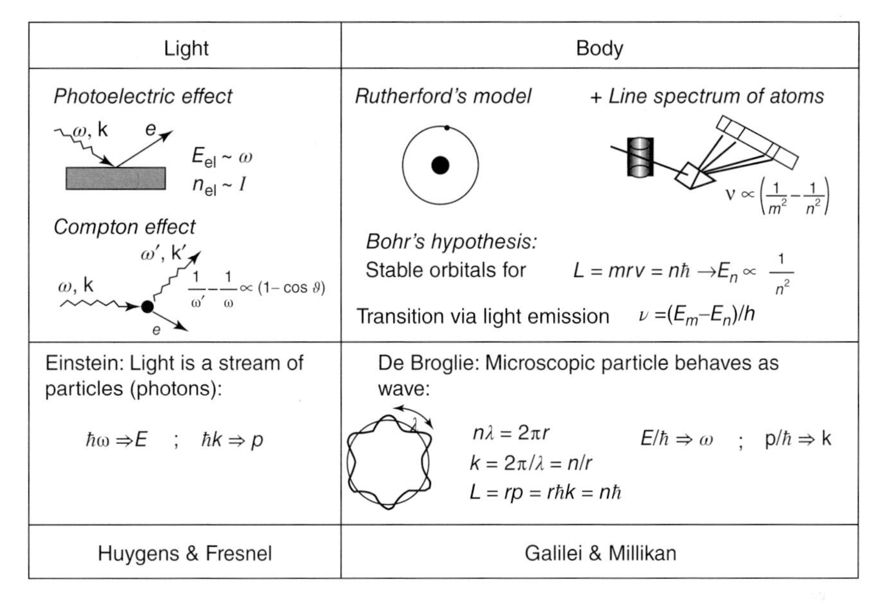

Figure 5.3 The duality problem of light and bodies.

else. In contrast, a wave with angular frequency ω and wave number k can be found everywhere with the same probability.

The dichotomy between point-mass-like particles and light waves was seriously disturbed by the realization of Planck that energy between them can only be exchanged in $h\nu$ quanta. Starting from there, it became clear that light can behave as a stream of point-mass-like particles (photons), and a stream of point-mass-like electrons can behave as a wave (Figure 5.3). The photon hypothesis of Einstein (to explain the photoelectric and Compton effects, and confirmed by the discovery of laser) was now contrasted with the interpretation of light interference in terms of waves, as introduced by Huygens and Fresnel. Similarly, the hypothesis of electron waves by de Broglie (which can explain the interpretation of the line spectrum of atoms in terms of the Bohr model) is in contrast with the whole experience of classical physics from Galilei to Millikan. This duality is, from the viewpoint of physics, not really acceptable.

5.2 Double-Slit Experiment with a Single Electron

To resolve the problem, the following question should be raised: *what kind of a wave is a single electron*? One should consider that it cannot be a material wave (propagating vibration of some material medium) because then it could be divided, contrary to the Millikan experiment. In the Davisson–Germer experiment, an electron beam (stream of electrons) was used. Applying sophisticated techniques (an advanced electron gun and a charge-coupled device, CCD, as detector plate), the behavior of a single electron can also be studied in

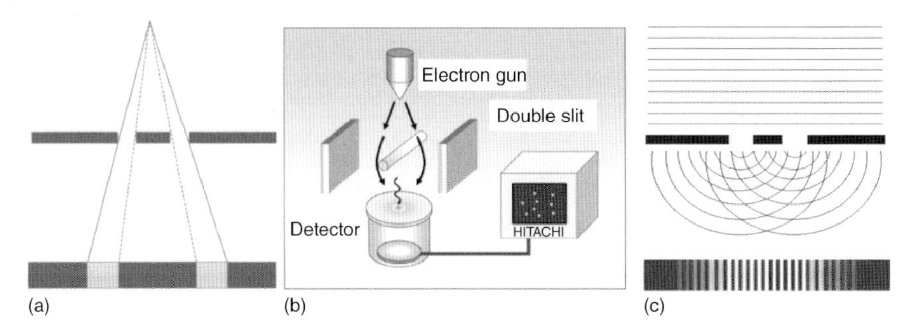

Figure 5.4 Expected outcome of a double-slit experiment (b) with a single electron according to the duality principle: classical point mass (a) and classical wave (c). (Reproduced with permission of Motoyasu Tsunoda, Central Research Laboratory, Hitachi Ltd.)

a double-slit experiment (Fig. 5.4b).[4] This is shown in a video clip prepared by the research institute of Hitachi (http://www.hitachi.com/rd/portal/research/em/doubleslit.html). If the electron was a point mass, the detector would find all the sequentially fired electrons in one of two areas behind the double slit (see the Figure 5.4a). If the electron was a wave, *each* of them ought to produce an interference pattern (Figure 5.4c), as light does on a double slit (Young's experiment, see Section A.6.1). As can be seen in Figure 5.5, none of these happens!

In accordance with the Millikan experiment and the point-mass concept, each electron gives rise to a small spot on the detector, that is, they seem to be localized. However, such spots appear also outside the classical "shadow zone" of the slits, where no point mass is supposed to arrive. After a couple of hundred shots, one can see that the impacts are randomly distributed. However, continuing the experiments for a sufficiently long time, a stripe pattern of the impacts emerges, corresponding exactly to the intensity distribution of light (Eq. (A.28)) in Young's double-slit experiment. Since the individual electrons were fired independent of each other, this interference pattern indicates that each electron must have realized the presence of both slits. That is only possible if – in agreement with the

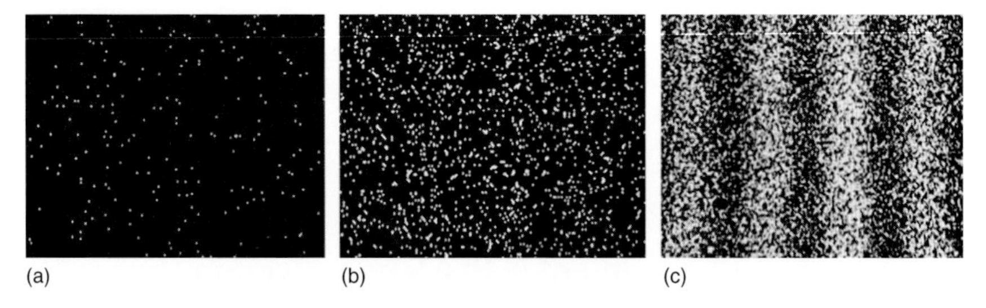

Figure 5.5 Actual outcome of the double-slit experiment with single electrons detected on a CCD screen. In (a–c) snapshots are shown after an increasing number of electron shots. The experiment was performed by Tonomura *et al.* [1]. (Reproduced with permission of Motoyasu Tsunoda, Central Research Laboratory, Hitachi Ltd.)

4 In fact, this experiment has been carried out without such modern tools but with unbelievable skill already in 1949 by *V. Fabrikant* in Moscow and in 1957 by C. Jönsson in Tübingen.

de Broglie hypothesis – the electron was a delocalized wave after leaving the gun, and was only localized by the detector in randomly distributed spots.[5]

5.3 The Born–Jordan Interpretation of the Electron Wave

The contradictions related to the wave nature of the electron were resolved by *Max Born* and *Ernst Pascual Jordan.* Following their statistical interpretation:

(a) The impact of the electrons is random and the detector shows the probability density for finding an electron in space (i.e., in the plane of the detector).
(b) Since the observed probability density corresponds to an interference pattern, it has to be proportional to the absolute square of a wave function $|\psi^*(\mathbf{r},t)\cdot\psi(\mathbf{r},t)|$.
(c) The wave function $\psi(\mathbf{r},t)$ itself cannot be measured (is not an observable). It is a complete mathematical description about the state of the particle and can be used as a source of information about its physical properties.

This last revolutionary assumption is the basis of quantum mechanics.

5.4 Heisenberg's Uncertainty Principle

As discussed in the previous subsection, the electron shows, on the one hand, wave-like behavior (interference) in the double-slit experiment but, on the other hand, the detector finds it more or less localized. One can guess, therefore, that the wave function of the electron must describe a wave packet of finite length (see Section A.5). Let us consider the relation of the phase and group velocities for this packet. By expanding the definition of the group velocity (Eq. (A.23)) by $\hbar$ and using the hypothesis of de Broglie in Eq. (4.11),

$$\mathrm{v}_g = \frac{d\omega}{dk} = \frac{d\hbar\omega}{d\hbar k} = \frac{dE}{dp} = \frac{d}{dp}\frac{p^2}{2m} = \frac{p}{m} = \mathrm{v} \tag{5.1}$$

So the wave packet moves with the speed of the classical (point-mass-like) particle. The phase velocity, however, is

$$\mathrm{v}_f = \frac{\omega}{k} = \frac{\hbar\omega}{\hbar k} = \frac{E}{p} = \frac{1}{p}\frac{p^2}{2m} = \frac{p}{m} = \frac{\mathrm{v}}{2} \neq \mathrm{v}_g \tag{5.2}$$

that is, different from the group velocity. This means that the wave packet, describing the electron, propagates dispersively and must delocalize with time. In other words, the electron cannot stay in a state similar to a point mass! Using

5 A word of warning is due here. This and similar conclusions of quantum mechanics are often interpreted as if observation would change the outcome of an experiment. One should, however, never forget that observation does not mean the mere presence of the observer. It means a measurement, that is, an interaction with the observed particle. We have seen in Chapter 3 that even the interaction with the lightest object (a photon) can change the state of an electron.

Eq. (A.24) for the width of a dispersive wave packet in space, after multiplication by h we get

$$\Delta x \cdot \Delta k > 2\pi \rightarrow \Delta x \cdot \Delta \hbar k > 2\pi\hbar \rightarrow \Delta x \cdot \Delta p > h \tag{5.3}$$

The spatial width of the packet can be regarded as the uncertainty of the position x, since the particle can be found with finite probability anywhere within Δx. Similarly, Δp is the uncertainty of the momentum. That is why Eq. (5.3), derived by *Werner Heisenberg*, is called the uncertainty principle. The product of the two uncertainties is larger than a constant, so if one is small, the other must be large. It means that *the position and momentum* (speed) *of an electron cannot be measured accurately at the same time* (i.e., in the same state of the particle). The closer the state is to the classical concept of the point mass ($\Delta x = 0$), the larger the uncertainty of the speed. However, if the speed is completely undetermined ($\Delta p = \infty$), the very concept of a trajectory becomes meaningless. In end effect, contrary to the particle concept of classical physics, an electron cannot be ever characterized with accurately measurable momentum and spatial position.

5.5 Particle Concept of Quantum Mechanics

We can formulate the uncertainty principle with the help of probabilities, as shown in Figure 5.6: a single value with probability 1 for position and momentum is not simultaneously possible. Subatomic particles cannot be point masses! Later, we will also see that the completely delocalized wave concept of the classical field theory is not possible either. Instead, electrons and other subatomic particles can have many states, some characterized by a more defined position and a less defined momentum, some vice versa, and some in between, as shown in Figure 5.7. The states with strong localization in x come closest to the classical point mass, and the ones with best defined p_x to the classical wave.

The possible forms of existence (states) of subatomic particles are more numerous than the ones we know in the macroscopic world; however, these states are difficult to comprehend. We can easily *imagine* the concept of the classical infinite wave by thinking of the *image* of a water wave and mentally extend it beyond the horizon. We can also *imagine* a point mass by thinking of the *image* of a small

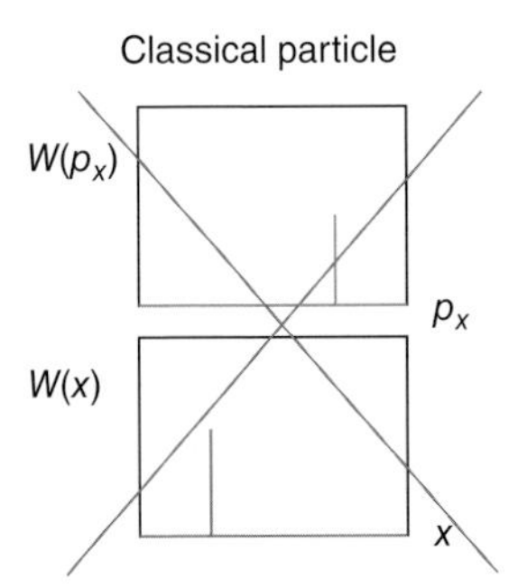

Figure 5.6 The point-mass concept of classical particles, expressed in terms of the probability of observing a given position and momentum. According to Eq. (5.3), this concept cannot be applied to an electron.

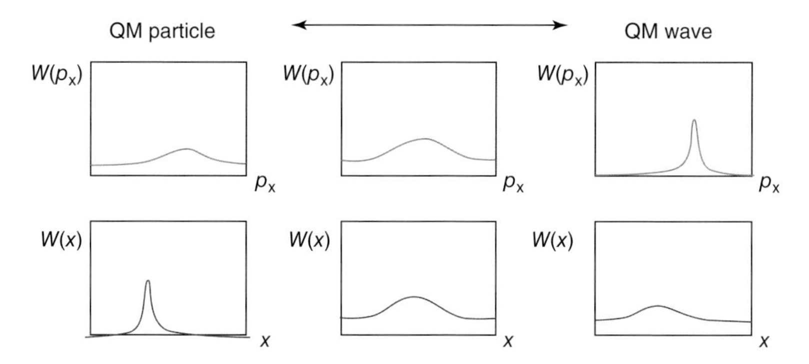

Figure 5.7 Possible states of an electron, expressed in terms of the probability of observing a given position and momentum.

pellet and mentally shrink it till we cannot see it. In other words, our understanding of the idealized classical concepts is based on images we observe in everyday life. However, we cannot follow this tactics with subatomic particles, because the states they can have just do not occur at all in our macroscopic world.

5.6 The Scale Dependence of Physics

Subatomic particles, such as electrons, necessitate a lot more complicated physics than the bodies we can actually observe. These complications disappear in the limit of large masses, similarly, as the consequences of the theory of special relativity become negligible for small velocities. As an analogy one can consider the following "Parable of the parabola people."[6]

Once upon a time there was a 1D universe with non-Euclidean geometry in the form of a parabola (Figure 5.8). In this universe have lived 1D people, somewhere in the upper range, where space was nearly a straight line. Actually, they could not even measure the deviation from linearity, so they have developed first Euclidean geometry and then built their physics on that. After a while, they set out to explore their world. At first, up the parabola, into the "Macroworld." Their measurements showed that their physics allowed even more accurate predictions

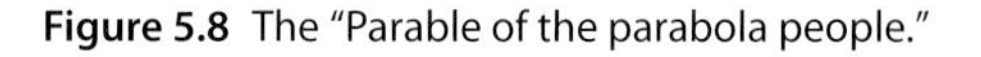

Figure 5.8 The "Parable of the parabola people."

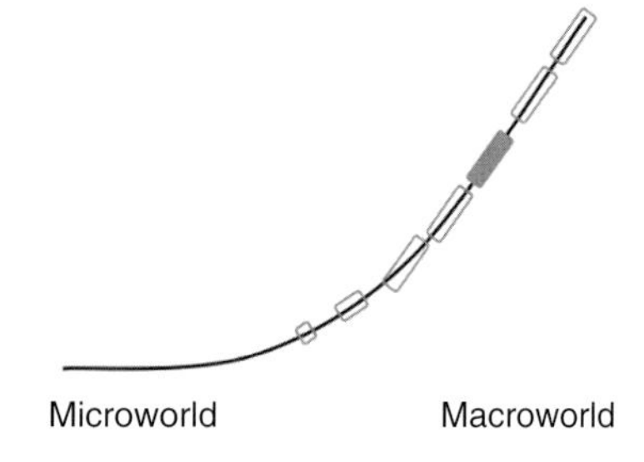

6 I have heard this tale from the renowned particle physicist, G. Marx, my unforgettable quantum mechanics teacher at the Lóránd Eötvös University, Budapest.

than in their everyday environment. (The higher up one goes on a parabola, the smaller the curvature. This experience is similar to ours when applying Newtonian mechanics to celestial bodies, where friction and interaction with other bodies becomes negligible.) Then they attempted to go in the other direction, to discover their "Microworld." They found, however, that they cannot go very far. Their body possessed a limited flexibility they were not even aware of. But one of them got the idea of sending down small probes. The smaller the probe was, the further it could travel. So, finally, they deduced – for them – the completely abstract concept of curvature, and worked out parabolic mathematics and physics, without ever being able to imagine curvature. Something similar to this happens with us, too, when we try to explore our "Microworld."

The lesson is that the laws of physics are not scale independent, as we have assumed them to be in classical physics. The assumption that any physical phenomenon can, in principle, be modeled in any size, has allowed us to choose arbitrarily the units of the basic observables: time, distance, and mass, as (s), (m), and (kg), respectively. We have, however, learnt from the special relativity theory that time multiplied by the speed of light, ct, should be handled in an equivalent manner with the spatial coordinates x, y, z (see Eq. (A.44)). This means that the units of time and distance are not independent. If one selects (s) for the former, the natural choice for the latter is the [light · second]. The general relativity states that the gravitational force is nothing but the curvature of space around a mass. Therefore, its unit (in classical physics $[\text{kg} \cdot \text{m}^2 \cdot \text{s}^{-2}]$), and with that the unit of mass, should not be chosen independently. Finally, Heisenberg's uncertainty principle in Eq. (5.3) established another relation between the basic observables, which means that we cannot freely choose the unit of any of them! Nature has its own system of units and, with our arbitrary choice, we have disregarded the scale dependence of its laws.

5.7 Toward a New Physics

Based on the interpretation of Born and Jordan, we must attempt to rebuild mechanics into a new *quantum mechanics*. First of all, we are going to need an equation to determine the wave function, which contains all the information about the state of a particle. We present here a heuristic derivation in advance, which shows that the required equation will be based – just like the Newtonian equation of motion – on knowledge of the potential energy, as a function of position and time.

The wave function $\psi(\mathbf{r}, t)$ should obviously satisfy a wave equation, similar to Eq. (A.17). We also know that any arbitrary wave can be mathematically constructed by a linear combination of harmonic waves (Fourier analysis). The wave equation contains only derivatives, that is, only linear operators (the constituents of a sum of functions can be differentiated independently), so let us consider only one element of the Fourier expansion:

$$\psi(x, t) = \psi_0 e^{i(kx-\omega t)} \tag{5.4}$$

and substitute it into the wave equation

$$\frac{\partial^2 \psi}{\partial t^2} = v_f^2 \frac{\partial^2 \psi}{\partial x^2} \tag{5.5}$$

After differentiating once on the left-hand side and twice on the right-hand side, we obtain

$$- i\omega \frac{\partial}{\partial t} \psi_0 e^{i(kx-\omega t)} = -v_f^2 k^2 \psi_0 e^{i(kx-\omega t)} \tag{5.6}$$

Using the dispersion relation $\omega = v_f \cdot k$ (Eq. (A.19)), as well as $\hbar\omega = E = T + V$, we obtain a differential equation (the so-called *Schrödinger equation*) for the determination of $\psi(x, t)$:

$$i\hbar \frac{\partial \psi}{\partial t} = (T + V)\psi \tag{5.7}$$

which requires knowledge of the potential energy V, just as the Newtonian equation of motion

$$m \frac{d^2 x}{dt} = -\frac{dV}{dx} \tag{5.8}$$

does. The difference is that in Newtonian mechanics the state of the particle is always assumed to be that of a point mass, and the potential energy determines only its trajectory. In quantum mechanics, the state of the particle itself (e.g., localized or delocalized) depends on V.

5.8 The Significance of Electron Waves for Electrical Engineering

Even though the quantum mechanical concept of an electron cannot be "imagined," it plays an important role in understanding electronics. For example, according to the well-known Ohm's law (Eq. (1.5)), the current ought to be proportional to the voltage. This is true in metal wires but the *I-V* characteristics of some semiconductors (e.g., GaAs) are similar to the one shown in Figure 5.9. An explanation of the sudden increase in resistivity can only be given in semiconductor physics based on the quantum mechanical nature of electrons. The wave-like behavior namely leads to the superposition of the many individual electrons in a solid. Although the arising wave packets can be regarded as relatively well-localized particles carrying an electron's charge, quantum mechanics shows that their effective mass will depend on their speed

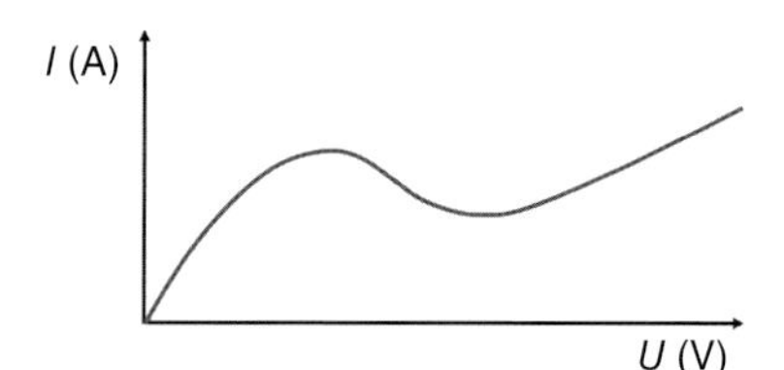

Figure 5.9 *I–V* characteristics of the semiconductor GaAs, used, for example, in Gunn diodes.

in a complicated manner.[7] A more detailed explanation cannot be supplied here, but the so-called *negative differential resistance*, shown in Figure 5.9, is the basis of *Gunn* diodes, used to provide the clock signal of computers.

5.9 Displaying Electron Waves

Even though there is no way to imagine the quantum mechanical state of an electron, some aspects of the state can be made visible by sophisticated electronic engineering. The *scanning tunneling microscope* (STM) utilizes the quantum mechanical tunnel effect, that is, the unimaginable ability of an electron to go through a potential barrier even if its kinetic energy is lower than the height of the barrier (similar to going through a wall without jumping over it or breaking through it). As discussed later in Chapter 10 in more detail, electrons of a metal surface can reach an electrode separated from the surface by vacuum, that is, by an isolating layer. The resultant current will depend exponentially on the thickness of the layer. Using a very sharp pin as electrode, and steering it over the surface by piezoelectric controllers, a current map can be recorded and displayed. Due to the high sensitivity to distance, even atomic resolution can be reached. Figure 5.10 shows a "fence" built by 48 iron atoms on a copper surface. Inside the fence, a single electron is captured, which, under these conditions (round potential well) can be described by a standing wave. This changes the potential of the surface according to the intensity of the standing wave, and that is observed in the STM image.

It should be emphasized that what we *see* here is not the electron itself but the modification of the tunnel current due to its presence. However, this modification corresponds to the wave nature of the electron, as predicted by quantum mechanics, and the experiment shows that the consequence of the quantum mechanical wave function is a measurable reality.

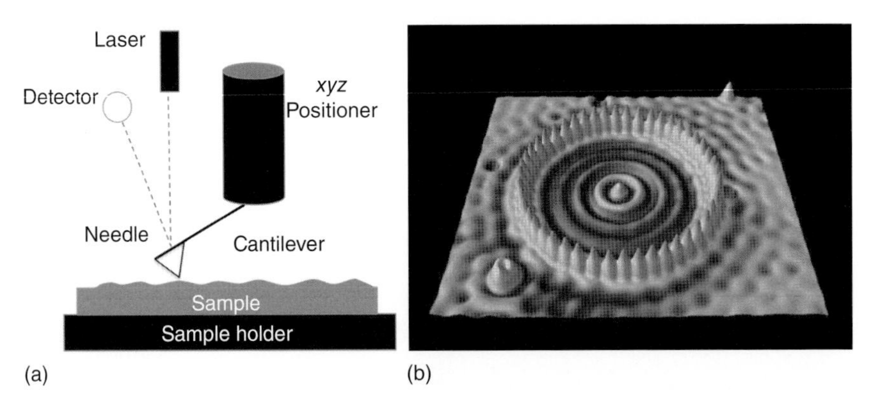

Figure 5.10 Schematic of a scanning tunneling microscope and the current image of an electron captured as standing wave within a fence of 48 iron atoms on a copper surface. Color online. (b: Reproduced with permission of Michael Crommie, Department of Physics, UC Berkeley.)

7 This has nothing to do with the relativistic increase of the mass with speed.

Summary in Short

- The double-slit experiment with a single electron shows that a subatomic particle should be regarded as a wave packet: it is capable of interference (as a wave should) but appears localized on the screen (as a body). (N.B.: We are talking about concepts here!)
- Based on the correspondence $E \leftrightarrow h\nu$; $p \leftrightarrow h/\lambda$ (de Broglie), the propagation of the wave packet must be dispersive. From that follows that the position and momentum cannot be measured accurately in the same state of the particle (Heisenberg's uncertainty principle).
- The electron (and other subatomic particles) can have many states. Some can be characterized with a definite position in space, some with a definite momentum, but there are states in which it makes no sense talking about either position or momentum. Position, momentum, and all other observables have been defined by macroscopic experiments. Subatomic particles have "life forms" different from macroscopic bodies. The laws of nature are not scale independent!
- Born–Jordan interpretation: the wave function ψ, which describes the state of the particle, is not an observable quantity (it is not a material wave). It is a complex function (in the mathematical sense), providing information about the state, for example, the absolute square of the wave function describes the probability density of finding the particle in space.
- The wave function can be determined with knowledge of the potential energy from the Schrödinger equation.
- **To be able to predict observable physical quantities** (which have been defined for macroscopic bodies in the first place) **for an electron** (or other subatomic particles), **the development of quantum mechanics is needed.**

5.10 Questions and Exercises

Problem 5.1 Describe the two main concepts of classical physics for the different forms of matter!

Problem 5.2 How can the double-slit experiment with single electrons be explained, based on the Born–Jordan interpretation of the concept of a subatomic particle?

Problem 5.3 What determines the state of a particle in quantum mechanics?

Problem 5.4 Considering the uncertainty principle of Heisenberg, what quantum mechanical states come closest to the classical concepts of a point mass and a wave?

Reference

1 Tonomura, A., Endo, J., Matsuda, T., Kawasaki, T., and Ezawa, H. (1989) *Am. J. Phys.*, **57** (2), 117–220.

6

Measurement in Quantum Mechanics. Postulates 1–3

In this chapter…
we start with the systematic construction of quantum mechanics. The laws of classical mechanics could be derived from a small number of axioms mathematically and were then confirmed by experiments. *Axioms* are rules that are established on the basis of countless observations and controlled experiments. They are acknowledged as the basic characteristics of reality, without trying to find an explanation for them. (The axioms tell how we perceive the world to be.) In quantum mechanics, we cannot rely on direct observation (with our senses), and, considering that we cannot even imagine the abstract concept of a particle, we can only formulate assumptions mathematically, based on a limited set of experiments. These assumptions are called *postulates*. Postulates are verified subsequently by experimental confirmation of the laws derived from them. (The postulates tell how we assume the world to be.) Similar to axioms, there is a certain freedom in choosing the postulates.[1] In addition, their formulation depends on the mathematical apparatus we want to use. In this book, we follow the so-called *Schrödinger picture* (based on differential equations[2]) and the line of thought of *P. A. M. Dirac.* In setting up the postulates, one should consider that *the electron* (or another subatomic particle) *is not an object* (which we can see or hold) *but a concept, which can only be formulated mathematically with the help of the wave function,* based on our knowledge gained from complicated experiments. The central question in this chapter is as follows: how does quantum mechanics make it possible to predict the outcome of measurements from knowledge of the wave function? (We discuss the determination of the wave function in Chapter 7.)

1 Like, for example, the choice between the Newtonian equation of motion, Eq. (A.9), and the Hamiltonian equations of motions, Eq. (A.40).
2 In contrast to the Heisenberg picture, based on matrix equations of infinite size.

Essential Quantum Mechanics for Electrical Engineers, First Edition. Peter Deák.

6.1 Physical Restrictions for the Wave Function of an Electron

Based on the Born–Jordan interpretation, we expect that the state of an electron will be completely described by the wave function $\psi(\boldsymbol{r}, t)$. Our basic knowledge about the (concept of an) electron should help us to restrict the possible mathematical functions, which may describe it. The following four assertions suffice to pick a given class of functions.

Physical assertion: The electron behaves deterministically, that is, under the same conditions it behaves always the same way.
Mathematical formulation: $\psi(\boldsymbol{r}, t)$ **must be a single-valued function of its variables.**

Physical assertion: The electron is undividable.
Mathematical formulation: $\psi(\boldsymbol{r}, t)$ **must be a continuous function.**

Physical assertion: The electron has wave properties, that is, the wave function must satisfy a wave equation.
Mathematical formulation: $\psi(\boldsymbol{r}, t)$ **must be twice differentiable** (or more precisely, the first derivative must be a continuous function).

Physical assertion: Since the probability of finding the electron at the time t in the *space d**r** around the position **r** should be* $|\psi(\boldsymbol{r}, t)|^2 \cdot d\boldsymbol{r}$, the integral of the latter over the entire space should give the value of 1. (The electron must be somewhere, after all…)
Mathematical formulation: $\psi(\boldsymbol{r}, t)$ **must be square-integrable.**[3]

These four conditions are actually valid for any subatomic particle. Their capability for interaction (electrical charge, gravitational charge or mass, etc.) and their spin (see later in Chapter 12) identify the particular particle. In mathematics, these four conditions define the class of the so-called *regular functions.* We can summarize all that as follows:

Postulate 1

The state of the electron can be described by a wave function $\psi(\boldsymbol{r}, t)$, which is mathematically regular.

N.B.: Postulate 1 actually makes two claims:

(a) *A function exists that provides all information about the state of a particle.*
(b) *This function is single-valued, continuous, twice differentiable, and square-integrable* (i.e., the particle is deterministic, undividable, shows wave-like behavior, and the probability density of finding it in space is given by the absolute square of the wave function).

---

3 This means that the integral $\int \psi^*(\boldsymbol{r}, t) \cdot \psi(\boldsymbol{r}, t) d\boldsymbol{r}$ over the whole space must have a finite value. Any square-integrable function, divided by the square root of this value, satisfies the fourth physical assertion.

In some way, this postulate plays a similar role to the first Newtonian axiom, which selects the system of reference in which the equation of motion (Second Newtonian axiom) is valid (see Section A.2). Postulate 1 of quantum mechanics selects the physically allowed states from the mathematical solutions of the Schrödinger equation (Eq. (5.7), which is going to be our Postulate 5, see later in Chapter 7).

6.2 Mathematical Definitions and Laws Related to the Wave Function

The integral of the product of a wave function and the complex conjugate of another will play an important role in the following. Therefore, we introduce here a few abbreviations and definitions:

Abbreviation: *scalar product of two functions*

$$\int \psi_i^*(\mathbf{r},t)\psi_j(\mathbf{r},t) = \langle \psi_i | \psi_j \rangle \tag{6.1}$$

Definition 1: two functions, ψ_i and ψ_j, are called *orthogonal* if

$$\langle \psi_i | \psi_j \rangle = 0 \tag{6.2}$$

Definition 2: a function ψ_i is called *normalized* if

$$\langle \psi_i | \psi_i \rangle = 1 \tag{6.3}$$

Definition 3: a system of functions $\{\psi_i\}$ is called orthonormal[4] if for any two elements

$$\langle \psi_i | \psi_j \rangle = \delta_{ij} \tag{6.4}$$

where the so-called *Kronecker delta* stands for

$$\delta_{ij} \equiv \begin{cases} 1 & \text{if } i = j \\ 0 & \text{if } i \neq j \end{cases} \tag{6.5}$$

It is important to know that a complete set of functions (see Section B.2) can be made *orthonormal,* that is, one can construct linear combinations of the elements that are all normalized and orthogonal to each other.

---

4 An example for an orthonormal system is the functions

$$\varphi_n(x) = \frac{1}{\sqrt{2\pi}} e^{inx}, \quad n = \ldots, -1, 0, 1, \ldots \text{ and } x \in [-\pi, +\pi]$$

because

$$\begin{aligned}
\langle \varphi_n \varphi_m \rangle &= \frac{1}{2\pi} \int_{-\pi}^{\pi} \varphi_n^*(x)\varphi_m(x)dx = \frac{1}{2\pi} \int_{-\pi}^{\pi} e^{i(m-n)x}\, dx \\
&= \frac{1}{2\pi} \left[\frac{e^{i(m-n)x}}{i(m-n)} \right]_{-\pi}^{\pi} = \frac{1}{(m-n)\pi} \frac{e^{i(m-n)\pi} - e^{-i(m-n)\pi}}{2i} \\
&= \frac{\sin(m-n)\pi}{(m-n)\pi} = \begin{cases} n = m \to 1 \\ n \neq m \to 0 \end{cases}
\end{aligned}$$

6.3 Mathematical Representation of the Measurement by Operators

Let O be any observable defined by macroscopic measurement in classical mechanics, that is, the letter O is representing here all measurable quantities, such as momentum p, energy E, and so on. Let us emphasize again that "observable" means a property of the particle that can be measured and not merely observed passively.

In classical mechanics, the state of the particle is always the point mass, residing at the time t at the position x, and so all other observables can be given as mathematical *functions* of these variables:

$$O(x, t)$$

This means that the measured values of O can be unambiguously and continuously assigned to points in space and time, and the measurement does not change the state of the particle.

In quantum mechanics, the state of the particle is given by the wave function $\psi(x, t)$ and even the most careful measurement (e.g., "shining" a single photon onto an electron) might change the state. That is why the mathematical representation of the measurement must be an *operator* (a prescription for mathematical operations, see Section B.3), which is able to change the wave function:

$$\hat{O}\psi(x, t)$$

Of course, the question arises: how does the operator $\hat{O}$ assign a value of the observable O to a given state, that is, how does it tell us what we are going to measure? It is known that the eigenvalue equation of the operator, that is, $\hat{O}\varphi_i = k_i\varphi_i$ (see Eq. (B.3)), assigns the numerical value k_i to the eigenfunction φ_i. If we can somehow make sure that the k_i eigenvalues of the operator $\hat{O}$ are real numbers, we could attempt to interpret them as the measurable values of the observable O. It can be mathematically proven that all eigenvalues of an operator are real if the operator satisfies the following condition for any two regular functions:[5]

$$\langle \hat{O}\psi_1 | \psi_2 \rangle = \langle \psi_1 | \hat{O}\psi_2 \rangle \tag{6.6}$$

Such operators are called *Hermitian* or self-adjoint operators.

In order to satisfy the superposition principle, operators describing a measurement must be linear (e.g., the differential operator but not the square root operator):

$$\begin{aligned} \hat{O}(f + g) &= \hat{O}f + \hat{O}g \\ \hat{O}(c \cdot f) &= c \cdot (\hat{O}f) \end{aligned} \tag{6.7}$$

where c is a number and f, g are functions. We can summarize all that as

5 N.B.: The angle brackets mean a scalar product according to Eq. (6.1), that is, an integral in which the first function occurs complex conjugated.

Postulate 2

The measurement of an observable O is represented by a linear Hermitian operator $\hat{O}$, the eigenvalues of which being the possible results of the measurement.

N.B.: Postulate 2 also makes two claims:

(a) *The possible results of a measurement are determined in advance by the operator.* Therefore, we will have to select a specific operator for the measurement of every observable: this is discussed in Chapter 7. Since in mathematics it is well known that some operators have a discrete eigenvalue spectrum, we will be able to account for the quantization of some observables.

(b) *The operators representing a measurement must be linear and Hermitian.* These conditions make sure that the superposition principle, a generally observed characteristics of nature, is observed, and the values we can measure are real numbers.

How the operators assign a specific value of the observable to the actual state of the particle is described in Section 6.5.

6.4 Mathematical Definitions and Laws Related to Operators

As in Section 6.2, we have to introduce first a few mathematical theorems and definitions concerning operators.

Theorem 1: The eigenfunctions of an operator constitute a complete set and can be made orthonormal. Therefore, from now on, we consider them to be orthonormal in the first place, that is,

$$\text{if} \quad \hat{O}\varphi_i = k_i\varphi_i, \text{ then } \langle\varphi_i|\varphi_j\rangle = \delta_{ij} \tag{6.8}$$

Definition 1: Sum of two linear Hermitian operators $\hat{A}$ and $\hat{B}$:

$$(\hat{A} + \hat{B})\varphi \equiv \hat{A}\varphi + \hat{B}\varphi$$

The addition of operators is commutative.

Definition 2: Product of two linear Hermitian operators $\hat{A}$ and $\hat{B}$:

$$(\hat{A}\hat{B})\varphi \equiv \hat{A}(\hat{B}\varphi)$$

The multiplication of operators is not always commutative, that is, applying the two operators in the reverse order may lead to a differing result: $(\hat{A}\hat{B} - \hat{B}\hat{A})\varphi \neq 0$. (As an example, one can consider raising $\varphi(x) = x^2$ to some power and then differentiate it.)

Abbreviation: *commutator*

$$[\hat{A}, \hat{B}] \equiv (\hat{A}\hat{B} - \hat{B}\hat{A}) \tag{6.9}$$

Theorem 2 (Von Neumann's theorem): Two operators, $\hat{A}$ and $\hat{B}$, can only have common eigenfunctions if $[\hat{A}, \hat{B}] = 0$. If the commutator is zero, they will have at least one.

6.5 Measurement in Quantum Mechanics

According to Postulate 2, the operator $\hat{O}$ assigns the eigenvalue k_n to the eigenfunction φ_n. But how can $\hat{O}$ be used to predict the value of an observable in a given state ψ? We should consider two cases.

(a) The wave function is an eigenfunction of $\hat{O}$, $\psi = \varphi_n$. **This is a so-called eigenstate of the observable, and the measurement will give k_n, that is, the eigenvalue assigned to the eigenstate by the operator.**
(b) The wave function is not an eigenfunction of $\hat{O}$, $\psi \notin \{\varphi_i\}$. **This is a so-called mixed state from the viewpoint of the observable,** because it can be expanded by the complete set of eigenfunctions, $\psi = \sum_i c_i \varphi_i$, **and the measurement may give any of the k_i eigenvalues.**

We assume here that the measurement does not change an eigenstate (because the operator does not change an eigenfunction). The result of the measurement on the same eigenstate φ_n will always be k_n. The eigenstate can be characterized by a definite value of the observable. In contrast, in a mixed state, the observable is not characteristic to the state, but the measurement forces the particle into an eigenstate. As a consequence, the measurement will yield the eigenvalue of the final state. However, the outcome is a probability event. Starting from one given mixed state, the measurement may result in any of the k_i eigenvalues (after forcing the particle into any of the φ_i eigenstates), but with different probability $W(k_i)$.

The expansion $\psi = \sum_i c_i \varphi_i$ means that the various φ_i eigenstates participate in the mixture with different c_i weights. It makes sense to assume then that $W(k_i)$ depends on c_i. The higher the weight of φ_i in the mixture (the more similar ψ is to φ_i), the more likely ψ will transform into φ_i. Let us consider, however, that for any probability event the sum of probabilities should be $\sum_i W(k_i) = 1$. It can be proven[6] that

$$\sum_i |c_i|^2 = 1 \tag{6.10}$$

Therefore, we may assume that $W(k_i) = |c_i|^2$. The absolute square of the coefficients can be determined[7] for any ψ if $\{\varphi_i\}$ is known (from solving the eigenvalue equation of $\hat{O}$):

$$W(k_i) = |c_i|^2 = |\langle \psi | \varphi_i \rangle|^2 \tag{6.11}$$

We can summarize all that as

6 According to Postulate 1, $\langle \psi | \psi \rangle = 1$, and, with the help of Eq. (6.8) follows

$$1 = \langle \psi | \psi \rangle = \left\langle \sum_i c_i \varphi_i \middle| \sum_j c_j \varphi_j \right\rangle = \sum_{ij} c_i^* c_j \langle \varphi_i | \varphi_j \rangle = \sum_{ij} c_i^* c_j \delta_{ij} = \sum_i |c_i|^2$$

7 Because $|\langle \psi | \varphi_n \rangle|^2 = \left| \left\langle \sum_i c_i \varphi_i \middle| \varphi_n \right\rangle \right|^2 = \left| \sum_i c_i^* \langle \varphi_i | \varphi_n \rangle \right|^2 = \left| \sum_i c_i^* \delta_{in} \right|^2 = |c_n|^2$.

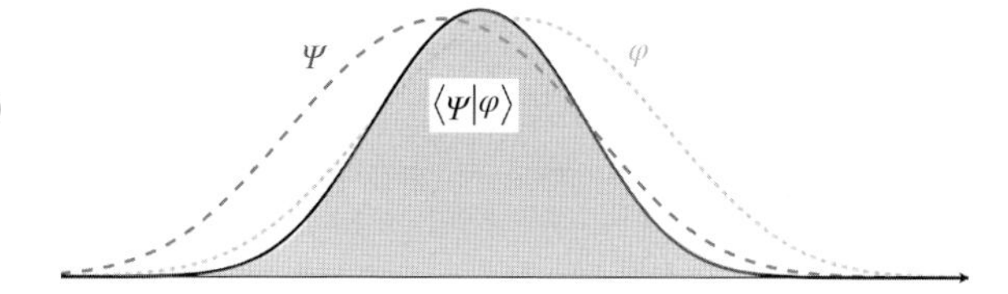

Figure 6.1 The meaning of the scalar product of ψ (dashed line) and φ_i (dotted) shown by the shaded area. (Both functions were assumed to be real here.)

Postulate 3

The transformation probability from an initial state ψ into the eigenstate φ_i of the operator $\hat{O}$ (due to the measurement of the observable O), and the probability of measuring the corresponding eigenvalue k_i, is $W(k_i) = |\langle\psi|\psi\rangle|^2$.

N.B.: Postulate 3 also makes two claims:

(a) *The measurement of one of the allowed values (eigenvalues of $\hat{O}$, see Postulate 2) is accompanied by the transformation of the measured state into the corresponding eigenstate.*
(b) *This transformation is a probability event. The larger the overlap of the original state ψ with the eigenfunction φ_i, the higher the probability for the transformation.* (The scalar product of two functions is actually the area below the function $\psi^*\varphi_i$, see Figure 6.1.)

As we can see Postulate 3, which is the measurement principle of quantum mechanics, says nothing about the outcome of a single measurement. It determines only the frequency with which a given value occurs as result in a large number of experiments. For example, in the double-slit experiment, any position on the detector plate can apparently be measured for the electron, and the probability distribution becomes visible only after a large number of measurements. Obviously, the electron passing the slits is not in an eigenstate of the spatial position, that is, it cannot be characterized by a definite position in space. The detector performs actually a measurement of position, forcing the electron into a new, localized state. The observed probability density witnesses that the initial state was an interference wave.

Different values obtained in several measurements on the same object are not at all unknown in the macroscopic world either. While *in classical physics we assume that the error of the measurement could, in principle, be eliminated, in actual fact it always happens.* Therefore, it is required – especially in engineering – to make a series of measurements and to consider the mean value (Eq. (B.5)) as result. The accuracy of the measurement is given by the standard deviation (Eq. (B.6)). *Postulate 3 of quantum mechanics tells us that in measurements on a mixed state the standard deviation from the mean value cannot be zero, as a matter of principle!*

Actually, in the case of the double-slit experiment, the mean value (middle of the screen) does not help us much, but let us consider two examples depicted in Figure 6.2. The dashed lines show hypothetical wave functions ψ, and the solid lines correspond to hypothetical eigenfunctions of the operator assigned to some observable. Due to the overlaps, the measurement in the first case will result in k_{-1} in nearly half of the cases and, just as frequently, in k_{+1}. The probability of

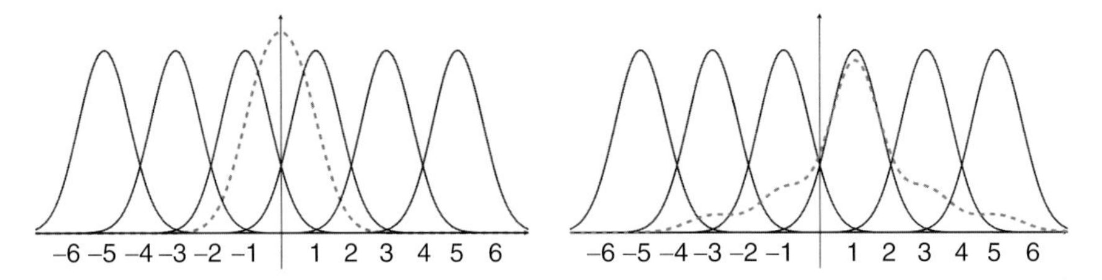

Figure 6.2 Relation of two states ψ (dashed) to the eigenfunctions φ_i (solid) of the operator representing the measurement ($i = -6, \ldots, 0, \ldots, +6$).

measuring any other eigenvalue is going to be minuscule. In the second case, k_1 will be measured most often but there will be some scattering of the results. These cases show that it makes sense mostly to calculate the mean value and standard deviation of a quantum mechanical measurement.

From the definition of the weighted average follows[8] the mean value of the measured eigenvalues

$$\overline{O} = \int \psi^* \hat{O}\psi \, dx = \langle \psi | \hat{O}\psi \rangle \equiv \langle O \rangle \tag{6.12}$$

which is called the **quantum mechanical expectation value of the observable,** $\langle O \rangle$. Similarly, from the statistical definition of the standard deviation, one obtains[9]

$$\sigma = \sqrt{\langle \psi | \hat{O}^2 \psi \rangle - \langle \psi | \hat{O}\psi \rangle^2} = \sqrt{\langle \hat{O}^2 \rangle - \langle \hat{O} \rangle^2} \equiv \Delta O \tag{6.13}$$

which is called the **quantum mechanical uncertainty**, ΔO.

---

8 From Eqs. (B.5) and (6.11) follows that

$$\begin{aligned}\overline{O} &= \lim_{N\to\infty} \sum_i k_i \frac{N_i}{N} = \sum_i k_i \left(\lim_{N\to\infty} \frac{N_i}{N} \right) = \sum_i k_i W(k_i) \\ &= \sum_i k_i c_i^* c_i = \sum_i \sum_j k_i c_j^* c_i \delta_{ij} = \sum_i \sum_j k_i c_j^* c_i \langle \varphi_j | \varphi_i \rangle = \int \left(\sum_j c_j^* \varphi_j^* \right) \left(\sum_i c_i k_i \varphi_i \right) dx \\ &= \int \psi^* \hat{O}\psi \, dx = \langle O \rangle\end{aligned}$$

where we have used the standard definition of probability $W(k_i) = \lim_{N\to\infty} (N_i/N)$, as well as Eq. (6.8). Note: $\hat{O}\psi = \hat{O}\sum_i c_i \varphi_i = \sum_i c_i \hat{O}\varphi_i = \sum_i c_i k_i \varphi_i$.

9 From Eq. (B.6)

$$\begin{aligned}\sigma &= \lim_{N\to\infty} \sqrt{\frac{1}{N} \sum_i N_i (k_i - \overline{O})^2} = \sqrt{\sum_i \lim_{N\to\infty} \frac{N_i}{N} (k_i - \overline{O})^2} = \sqrt{\sum_i W(k_i)(k_i - \overline{O})^2} \\ &= \sqrt{\underbrace{\sum_i k_i^2 W(k_i)}_{\langle \psi | \hat{O}^2 \psi \rangle} - 2\overline{O} \underbrace{\sum_i k_i W(k_i)}_{\langle \psi | \hat{O}\psi \rangle} + \overline{O}^2 \underbrace{\sum_i W(k_i)}_{1}} \\ &= \sqrt{\langle \psi \hat{O}^2 \psi \rangle - \langle \psi | \hat{O}\psi \rangle^2}\end{aligned}$$

Quantum mechanics tells us that – in the general case – the result of a single measurement cannot be predicted, only the expectation value and the uncertainty (the mean value and the standard deviation) of many measurements. However, *if the initial state was an eigenstate of the observable to be measured, that is,* $\psi = \varphi_n$, *then we obtain*

$$\langle O \rangle = \langle \varphi_i | \hat{O} \varphi_i \rangle = \langle \varphi_i | k_i \varphi_i \rangle = k_i \langle \varphi_i | \varphi_i \rangle = k_i$$

$$\Delta O = \sqrt{\langle \varphi_i | \hat{O}(\hat{O}\varphi_i) \rangle - \langle \varphi_i | \hat{O}\varphi_i \rangle^2} = \sqrt{\langle \varphi_i | k_i (k_i \varphi_i) \rangle - \langle \varphi_i | k_i \varphi_i \rangle^2} = 0$$

that is, *all measurements on the same initial state will result in the same eigenvalue, without any scattering.*

The application of quantum mechanics means, therefore, the calculation of expectation values by Eq. (6.12) and uncertainties by Eq. (6.13). These can then be compared with the results of a series of measurements (on the same initial state). The equations have been derived from Postulate 3 and that is why the latter is called the measurement principle of quantum mechanics, even if the postulate itself refers only to probabilities.

The uncertainty of the measurement on a mixed state is not in contradiction with the deterministic nature of the particle. The uncertainty is big only if we attempt to measure an observable that is not characteristic (does not make sense) for the given state.[10] The observables of classical physics have been defined by measurements on macroscopic bodies. A subatomic particle can have many states, which cannot be characterized by all (sometimes even any) of these quantities. If we attempt, nevertheless, to measure, we force the particle into a state where this observable makes sense.

Unfortunately, we must accept the fact that quantum mechanics cannot describe the interaction of the measuring apparatus ("observer") with the object of the measurement. The way how the initial mixed state is transformed into the detected eigenstate remains unknown to us. For some physicists (or rather, for the philosopher in some physicists) this is not acceptable. That is the reason for the many attempts to develop a wholly causal quantum mechanics, in which the state of the particle can be unambiguously described with the help of hidden (not measurable) parameters, independent of the observer.[11] For example, in the so-called de Broglie–Bohm theory, the position is considered to be such a parameter and can, in principle, be determined from an additional equation (besides the Schrödinger equation). Unfortunately, however, the boundary conditions to solve this equation are, as a matter of principle, not known! One can say in general that all theories, going beyond the determination of the expectation value and uncertainty of the measurement on a classically inconceivable mixed state, are only unprovable attempts of interpretation after all, and – in the best case – provide the same expectation value and uncertainty at the end of the day as the quantum mechanics described here.

10 An analogy: one should not ask after the frequency of white light!

11 The amateur philosopher in the author of this book thinks that partitioning the world into separate concepts (like a single electron), independent of others (like the "observer"), is arbitrary and a rough approximation. This approximation allows us to use a practically applicable mathematical apparatus, but we have to pay the price of resigning the possibility for a complete description. By the way: Bell's inequalities prove that there can be no hidden parameters.

Summary in Short

- The particle concept of quantum mechanics:
  Postulate 1. The state of the electron can be described by a wave function $\psi(r, t)$, which is mathematically regular.
- Measurement principle of quantum mechanics:
  Postulate 2. The measurement of an observable O is represented by a linear Hermitian operator $\hat{O}$, the eigenvalues of which being the possible results of the measurement.
  Postulate 3. The transformation probability from an initial state ψ into the eigenstate φ_i of the operator $\hat{O}$ (due to the measurement of the observable O), and the probability of measuring the corresponding eigenvalue k_i, is $W(k_i) = |\langle\psi|\psi\rangle|^2$.
- How does quantum mechanics make it possible to predict the outcome of experiments? The way of elucidating measurable information from the wave function ψ is to calculate the quantum mechanical expectation value and uncertainty

  $$\langle O\rangle = \langle\psi|\hat{O}\psi\rangle; \quad \Delta O = \sqrt{\langle\hat{O}^2\rangle - \langle\hat{O}\rangle^2}$$

  where we assumed that the operator representing the measurement is known. The following chapter is devoted to this question.

6.6 Questions and Exercises

Problem 6.1 The normalized wave function of an electron, ψ, can be written as a linear combination of two normalized eigenfunctions, φ_1 and φ_2, of the operator $\hat{O}$ assigned to the measurement : $\psi = a(5\varphi_1 + 12\varphi_2)$. The eigenvalues corresponding to these eigenfunctions are k_1 and k_2, respectively.

a) Determine the real number *a*!
b) What are the possible outcomes of the measurement?
c) What are the probabilities of these results?
d) What is the expectation value of this measurement?

Problem 6.2 The repeated measurement of the energy of a classical particle as a function of time (under the same conditions) resulted in the values shown in the following table:

Time	Measurement 1	Measurement 2	Measurement 3
t_1	1.0	1.1	0.9
t_2	2.0	1.9	1.9
t_3	3.0	3.2	3.0

Plot the mean value of the measurements as a function of time and provide error bars corresponding to the standard deviation, $\pm\sigma$. Round up results to two decimal digits and use Eqs (B.5) and (B.6)!

Problem 6.3 When is a particle in an eigenstate and when in a mixed state from the viewpoint of the observable O? What is the prediction of quantum mechanics regarding the result of the measurement in these cases?

Problem 6.4 What are the consequences of the von Neumann theorem for the unambiguous characterization of a state with two observables A and B?

7

Observables in Quantum Mechanics. Postulates 4 and 5. The Relation of Classical and Quantum Mechanics

In this chapter…
We are going to select the operators for the observables (i.e., for the quantities defined by macroscopic measurements in classical physics), in accordance with the uncertainty principle of Heisenberg. The choice of the energy operator will lead us directly to the Schrödinger equation, which can be used for the determination of the wave function. We will also see that the relations found between the observables in classical mechanics remain valid between their operators (or for the time derivatives of their expectation values) and that classical mechanics will prove to be a borderline case of quantum mechanics.

7.1 The Canonical Commutation Relations of Heisenberg

According to Postulate 2, linear and Hermitian operators must be chosen to represent the measurement of an observable. Before actually choosing them, however, let us consider the consequences of Heisenberg's uncertainty principle, which we have formulated (in an approximate manner) in Eq. (5.3) as $\Delta p \cdot \Delta x \geq h$. This equation has been derived from Eq. (A.24), concerning finite wave packets. Using Eq. (A.16) for vibrations, which are finite in time, a similar relation can be derived between the uncertainties of the time and energy measurement:

$$\Delta t \cdot \Delta \omega \approx 2\pi \rightarrow \Delta t \cdot \Delta \hbar\omega \approx 2\pi\hbar \rightarrow \Delta t \cdot \Delta E \approx h \tag{7.1}$$

As is the case for position and momentum, time and energy cannot be measured accurately in the same state either. These two pairs are canonical conjugates (see Section A.8), so Heisenberg has made a generalization. He assumed that no canonically conjugated pairs (i.e., a coordinate-like observable q and the corresponding momentum-like observable $p = [\partial(T - V)]/\partial\dot{q}$) can be measured simultaneously with zero uncertainty. This means that no state can exist that would be an eigenfunction of both operators, $\hat{q}$ and $\hat{p}$. The *von Neumann* theorem (Section 6.4) tells us then that the corresponding operators cannot be commutative. This has to be taken into account in the choice of the operators, making sure

Essential Quantum Mechanics for Electrical Engineers, First Edition. Peter Deák.

that the choice leads to a generalized[1] uncertainty relation similar to Eqs (5.3) and (7.1). Heisenberg has studied the line spectrum of atoms and found an accurate relation between the Δt lifetime of the excited state and the $\Delta\omega = \Delta E/\hbar$ width of the spectrum line, which was emitted when the atom returned into the ground state. To be able to reproduce this relation,[2] he suggested

Postulate 4

Operators representing the measurement of canonically conjugated observables must fulfill the commutation relation:

$$[\hat{p}, \hat{q}] = \frac{\hbar}{i} \tag{7.2}$$

The commutator in Eq. (7.2) means $\hat{p}(\hat{q}\varphi) - \hat{q}(\hat{p}\varphi) = \frac{\hbar}{i}\varphi$. Using this, as well as Eqs (6.12) and (6.13), Heisenberg was able to derive the observed relation between the standard deviations of canonically conjugated observables:

$$\Delta p \Delta q \geq \frac{\hbar}{2} \tag{7.3}$$

which is *the accurate and general form of Heisenberg's uncertainty principle.*

7.2 The Choice of Operators by Schrödinger

Now, we have three conditions to be considered when picking the operators. According to Eqs (6.6), (6.7), and (7.2), the operators should be linear, Hermitian, and, in the case of canonically conjugated pairs, they should comply with the commutation relation of Heisenberg. *Erwin Schrödinger* suggested the following choice:

$$\hat{q} =: q\cdot; \;\; \hat{p} =: \frac{\hbar}{i}\frac{\partial}{\partial q} \tag{7.4}$$

The operator of a generalized coordinate q (e.g., x, y, z, or the azimuth angle ϕ) should be the multiplication with the coordinate, and the operator of its canonically conjugated generalized momentum (i.e., p_x to x, or L_z to ϕ) should be the derivation with respect to the coordinate, times $\hbar/i$. In the case of vector operators $\hat{\mathbf{r}}$ and $\hat{\mathbf{p}}$, this implies the following:

$$\hat{\mathbf{r}} =: \mathbf{r}\cdot \rightarrow \left\{ \begin{matrix} \hat{x} =: x\cdot & \hat{p}_x =: \dfrac{\hbar}{i}\dfrac{\partial}{\partial x} \\ \hat{y} =: y\cdot & \hat{p}_y =: \dfrac{\hbar}{i}\dfrac{\partial}{\partial y} \\ \hat{z} =: z\cdot & \hat{p}_z =: \dfrac{\hbar}{i}\dfrac{\partial}{\partial z} \end{matrix} \right\} \rightarrow \hat{\mathbf{p}} =: \frac{\hbar}{i}\nabla \tag{7.5}$$

---

1 Equations (5.3) and (7.1) have been derived from Eqs. (A.16) and (A.24). The latter are valid for the special case of localization in the form of $\sin(x\,\Delta k)/(x\,\Delta k)$ and $\sin(t\,\Delta\omega)/(t\,\Delta\omega)$, respectively, assuming the function values to be negligible beyond the first nodes of the functions at $\pm\pi$.

2 It should be noted that actually there is no operator for time measurement in quantum mechanics. Nevertheless, this was the way how Postulate 4 was found.

Schrödinger's operators (multiplication and differentiation) are obviously linear, and one can easily prove that they are Hermitian as well (i.e., $\langle \hat{O}\psi_1 | \psi_2 \rangle = \langle \psi_1 | \hat{O}\psi_2 \rangle$)[3] and comply with Postulate 4.[4]

N.B.: the operators of Schrödinger represent only one possible choice and that is why Eq. (7.4) is not considered a postulate.

7.3 Vector Operator of the Angular Momentum

Schrödinger's choice also applies to the azimuth angle ϕ and its canonically conjugated pair, the z-component of the angular momentum, L_z:

$$\hat{\phi} =: \phi\cdot\ ;\ \hat{L}_z =: \frac{\hbar}{i}\frac{\partial}{\partial \phi} \tag{7.6}$$

In order to find the operators for the other components of the angular momentum, let us rewrite $\hat{L}_z$ into Cartesian coordinates[5]:

$$\begin{aligned} \hat{L}_z \varphi(x, y, z) &= \frac{\hbar}{i}\frac{\partial \varphi}{\partial \phi} = \frac{\hbar}{i}\left(\frac{\partial \varphi}{\partial x}\frac{\partial x}{\partial \phi} + \frac{\partial \varphi}{\partial y}\frac{\partial y}{\partial \phi} + \frac{\partial \varphi}{\partial z}\frac{\partial z}{\partial \phi}\right) \\ &= \frac{\hbar}{i}\left(-y\frac{\partial \varphi}{\partial x} + x\frac{\partial \varphi}{\partial y} + 0\right) \end{aligned} \tag{7.7}$$

The result corresponds to the z-component of the following vector product:

$$\mathbf{r} \times \frac{\hbar}{i}\nabla = \begin{vmatrix} \mathbf{e}_x & \mathbf{e}_y & \mathbf{e}_z \\ x & y & z \\ \frac{\hbar}{i}\frac{\partial}{\partial x} & \frac{\hbar}{i}\frac{\partial}{\partial y} & \frac{\hbar}{i}\frac{\partial}{\partial z} \end{vmatrix} \tag{7.8}$$

where the calculation of the z-component has been highlighted (see Eqs (B.26) and (B.27)). Comparing to Eq. (7.5), it can be recognized that Eq. (7.8) is the vector product of the operators for position and momentum, that is,

$$\hat{\mathbf{L}} = \hat{\mathbf{r}} \times \hat{\mathbf{p}} \tag{7.9}$$

and we can calculate the other components of the angular momentum operator from Eq. (7.8). As we can see, *the operator of the angular momentum obeys the same mathematical relation to the operators of position and momentum, as the corresponding observables in classical mechanics.* We expect this for other observables as well.

---

3 $\int (x\varphi_1)^* \varphi_2\, dx = \int x\varphi_1^* \varphi_2\, dx = \int \varphi_1^* x\varphi_2\, dx = \int \varphi_1^* (x\varphi_2)\, dx$

$\int \left(\frac{\hbar}{i}\frac{\partial \varphi_1}{\partial x}\right)^* \varphi_2\, dx = \int -\frac{\hbar}{i}\frac{\partial \varphi_1^*}{\partial x}\varphi_2\, dx = \underbrace{\left[-\frac{\hbar}{i}\varphi_1^*\varphi_2\right]_{-\infty}^{\infty}}_{0} + \int \varphi_1^* \frac{\hbar}{i}\frac{\partial \varphi_2}{\partial x}\, dx$ where it was taken into account that x is a real number, and regular functions $\varphi(x)$ have to vanish in $x = \pm\infty$, in order to be square-integrable.

4 $\frac{\hbar}{i}\frac{\partial}{\partial x}(x\varphi) - x\frac{\hbar}{i}\frac{\partial \varphi}{\partial x} = \frac{\hbar}{i}\varphi + x\frac{\hbar}{i}\frac{\partial \varphi}{\partial x} - x\frac{\hbar}{i}\frac{\partial \varphi}{\partial x} = \frac{\hbar}{i}\varphi.$

5 One should consider that $x = r\cdot \sin\vartheta \cdot \cos\phi$; $y = r\cdot \sin\vartheta \cdot \sin\phi$; $z = r\cdot \cos\vartheta$, and, assuming the function upon which $\hat{L}_z$ acts, to be expressed as $\varphi(x, y, z)$, one can apply the chain rule of derivation, Eq. (B.14).

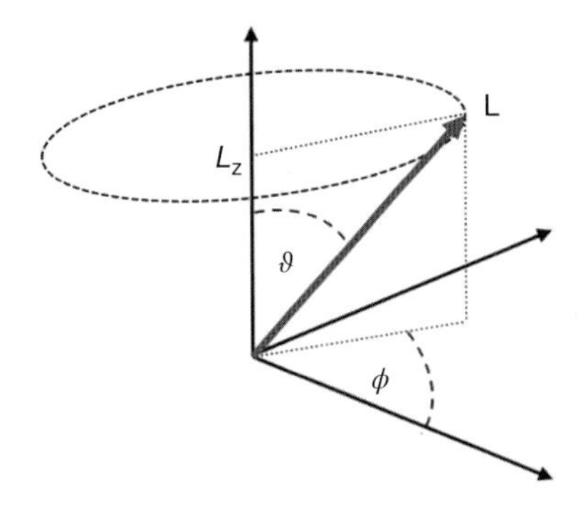

Figure 7.1 The angular momentum in polar coordinates.

Note that the right-hand side of Eq. (7.9) contains two operators that, according to Eq. (7.2), are not commutative. This means that all three components of the angular momentum vector cannot be determined accurately in the same state. It can be shown[6] that

$$[\hat{L}_i, \hat{L}_j] = i\hbar\hat{L}_k; \quad [\hat{L}_i, \hat{L}^2] = 0 \tag{7.10}$$

where i, j, and k are cyclic permutations of x, y, and z. This means that, with the exception of $\mathbf{L} = 0$, any other state can only be an eigenstate of one component of the angular momentum operator, but not of the others. If, say, the particle is in an eigenstate of $\hat{L}_z$, we can measure the magnitude (absolute value) of the vector $|\mathbf{L}|$ and the polar angle ϑ (i.e., L_z) with zero uncertainty, but not the azimuth angle ϕ (cf. Figure 7.1). After all, $\hat{L}_z$ and ϕ are canonically conjugated and must fulfill the uncertainty principle Eq. (7.3). This is a clear case where the world of subatomic particles cannot be squeezed into the simplified view of classical physics.

From this it follows that, on a circular orbit, one can know $|\mathbf{L}| = rmv$ and the orbital plane but not the point on the circumference where the particle is located. An electron cannot be localized along Bohr's orbitals as a point mass!

7.4 Energy Operators and the Schrödinger Equation

(Introduction of the Schrödinger equation following *K. Novobáczky*.) Schrödinger's choice for the operators defines also the operator of energy, as

$$\hat{E} =: \ -\frac{\hbar}{i}\frac{\partial}{\partial t} \tag{7.11}$$

However, the total energy in classical physics is defined by the *Hamilton function* (Eq. (A.41)) as $H = T + V$. The operator of the potential energy can be given with

6 For example,

$$\begin{aligned}
[\hat{L}_x, \hat{L}_y] &= (\hat{y}\hat{p}_z - \hat{z}\hat{p}_y)(\hat{z}\hat{p}_x - \hat{x}\hat{p}_z) - (\hat{z}\hat{p}_x - \hat{x}\hat{p}_z)(\hat{y}\hat{p}_z - \hat{z}\hat{p}_y) \\
&= \underline{\hat{y}\hat{p}_z\hat{z}\hat{p}_x} - \hat{y}\hat{p}_z\hat{x}\hat{p}_z - \hat{z}\hat{p}_y\hat{z}\hat{p}_x + \underline{\hat{z}\hat{p}_y\hat{x}\hat{p}_z} - \underline{\hat{z}\hat{p}_x\hat{y}\hat{p}_z} + \hat{z}\hat{p}_x\hat{z}\hat{p}_y + \hat{x}\hat{p}_z\hat{y}\hat{p}_z - \underline{\hat{x}\hat{p}_z\hat{z}\hat{p}_y} \\
&= \hat{y}\hat{p}_z\hat{z}\hat{p}_x - \hat{x}\hat{p}_z\hat{z}\hat{p}_y + \hat{z}\hat{p}_y\hat{x}\hat{p}_z - \hat{z}\hat{p}_x\hat{y}\hat{p}_z = -\hat{p}_z\hat{z}(\hat{x}\hat{p}_y - \hat{y}\hat{p}_x) + \hat{z}\hat{p}_z(\hat{x}\hat{p}_y - \hat{y}\hat{p}_x) \\
&= -\underbrace{(\hat{p}_z\hat{z} - \hat{z}\hat{p}_z)}_{[\hat{p}_z,\hat{z}]=\hbar/i}\underbrace{(\hat{x}\hat{p}_y - \hat{y}\hat{p}_x)}_{\hat{L}_z}
\end{aligned}$$

the help of the coordinate operators. For example, in the case of harmonic potential of the spring force, $V = 0.5Dx^2$, we could write $\hat{V} = 0.5D\hat{x}^2 = 0.5Dx\cdot(x\cdot)$. In general, the operator of the potential energy should then be

$$\hat{V} = V(\mathbf{r}, t)\cdot \tag{7.12}$$

According to Eq. (A.8), the kinetic energy is defined as $T = p^2/2m$. Using the operator of the momentum p, the operator of the kinetic energy should read

$$\hat{T} = \frac{\hat{\mathbf{p}}^2}{2m} = \frac{1}{2m}\frac{\hbar}{i}\nabla\left(\frac{\hbar}{i}\nabla\right) = -\frac{\hbar^2}{2m}\Delta \tag{7.13}$$

where we have used Eq. (B.9). (In one dimension, we would have $\hat{T} = -(\hbar^2/2m)(\partial^2/\partial x^2)$.) Using Eqs. (7.12) and (7.13), we can define the *Hamiltonian operator* of the energy as

$$\hat{H} = \hat{T} + \hat{V} = -\frac{\hbar^2}{2m}\Delta + V(\mathbf{r}, t)\cdot \tag{7.14}$$

Since we expect the mathematical relation between the observables of classical physics to be retained between their operators, and since we cannot have two different operators for the same observable, we require that the operators in Eqs (7.11) and (7.14) should lead to the same result:

$$-\frac{\hbar}{i}\frac{\partial\psi(\mathbf{r}, t)}{\partial t} = -\frac{\hbar^2}{2m}\Delta\psi(\mathbf{r}, t) + V\cdot\psi(\mathbf{r}, t) \tag{7.15}$$

or

$$-\frac{\hbar}{i}\frac{\partial\psi(\mathbf{r}, t)}{\partial t} = \hat{H}\psi(\mathbf{r}, t) \tag{7.16}$$

for any wave function ψ. These equations correspond to the Schrödinger equation of Eq. (5.7) and can be regarded as

Postulate 5

The wave function must fulfill the (time-dependent) Schrödinger equation $i\hbar\partial\psi/\partial t = \hat{H}\psi$.

N.B.: Equation (7.15) is a partial differential equation of second order and can be used to determine the wave function if the potential energy and the boundary conditions are known. It plays a similar role in quantum mechanics as Newton's second axiom, Eq. (5.8), does in classical mechanics. (The latter is an ordinary differential equation of second order, allowing the determination of the trajectory if the potential energy is known.) It is important to emphasize that a partial differential equation, such as the Schrödinger equation, can have many mathematical solutions. However, Postulate 1 tells us that we must accept only regular functions as solutions because only these can describe the state of a particle.

7.5 Time Evolution of Observables

Equations (7.9), (7.12), and (7.13) have shown that the classical relations of the observables are retained between the operators in quantum mechanics.

However, in classical mechanics, momentum and position are also related to each other by

$$p_x = m\frac{dx}{dt} \tag{7.17}$$

A similar equation cannot be given between the operators because (within the Schrödinger picture) the time derivative of an operator cannot be defined. We can, however, calculate the rate of change in the expectation value of the observable O:

$$\frac{d\langle O\rangle}{dt} = \frac{d}{dt}\langle\psi|\hat{O}\psi\rangle = \int\left[\frac{\partial\psi^*}{\partial t}\hat{O}\psi + \psi^*\hat{O}\frac{\partial}{\partial t}\psi\right]d\mathbf{r} \tag{7.18}$$

where we have used the derivation rule Eq. (B.12) and the definition of the scalar product, Eq. (6.1).[7] With the help of the Schrödinger equation follows[8]

$$\frac{d\langle O\rangle}{dt} = \frac{i}{\hbar}\int[(\hat{H}\psi^*)\hat{O}\psi - \psi^*\hat{O}(\hat{H}\psi)]d\mathbf{r} = \frac{i}{\hbar}\int[\psi^*(\hat{H}\hat{O} - \hat{O}\hat{H})\psi]d\mathbf{r} \tag{7.19}$$

where we have used Eq. (6.6) and the fact that $\hat{H}$ is Hermitian.

The integral on the right side is the expectation value of the commutator of $\hat{H}$ and $\hat{O}$, multiplied by $i/\hbar$. So, the *quantum mechanical time derivative* can be calculated as

$$\frac{d\langle O\rangle}{dt} = \frac{i}{\hbar}\langle[\hat{H},\hat{O}]\rangle \tag{7.20}$$

With the help of this general expression, we can also calculate the time derivative of $<x>$. Using Eq. (7.14)

$$\begin{aligned}\frac{d}{dt}\langle\hat{x}\rangle &= \frac{i}{\hbar}\langle\hat{H}\hat{x} - \hat{x}\hat{H}\rangle = \frac{i}{\hbar}\left(\frac{1}{2m}\langle\hat{p}_x^2\hat{x} - \hat{x}\hat{p}_x^2\rangle + \langle V\cdot x\cdot - x\cdot V\cdot\rangle\right)\\ &= \frac{i}{2m\hbar}\langle\hat{p}_x^2\hat{x} - \hat{x}\hat{p}_x^2\rangle\end{aligned} \tag{7.21}$$

where we have taken into account that the operators $V(x,t)\cdot$ and $x\cdot$ are commutative. One can show[9] that

$$\hat{p}_x^2\hat{x} - \hat{x}\hat{p}_x^2 = 2\frac{\hbar}{i}\hat{p}_x \tag{7.22}$$

and from that follows

$$m\frac{d\langle x\rangle}{dt} = \langle p_x\rangle \tag{7.23}$$

7 N.B.: The ordinary derivative with respect to t changes to a partial derivative, when applied to the function $\psi(\mathbf{r}, t)$.

8 If ψ is a solution of Eq. (7.16), so must be ψ^*: $\begin{matrix}\frac{\partial\psi}{\partial t} = -\frac{i}{\hbar}H\psi\\ \frac{\partial\psi^*}{\partial t} = +\frac{i}{\hbar}H\psi^*\end{matrix}$ and we can substitute these for the time derivatives in Eq. (7.18).

9 Using $\hat{p}_x\hat{x} - \hat{x}\hat{p}_x = \frac{\hbar}{i}$ it follows that $2\frac{\hbar}{i}\hat{p}_x = \hat{p}_x\frac{\hbar}{i} + \frac{\hbar}{i}\hat{p}_x = \hat{p}_x(\hat{p}_x\hat{x} - \hat{x}\hat{p}_x) + (\hat{p}_x\hat{x} - \hat{x}\hat{p}_x)\hat{p}_x = \hat{p}_x\hat{p}_x\hat{x} - \hat{p}_x\hat{x}\hat{p}_x + \hat{p}_x\hat{x}\hat{p}_x - \hat{x}\hat{p}_x\hat{p}_x = \hat{p}_x^2\hat{x} - \hat{x}\hat{p}_x^2$.

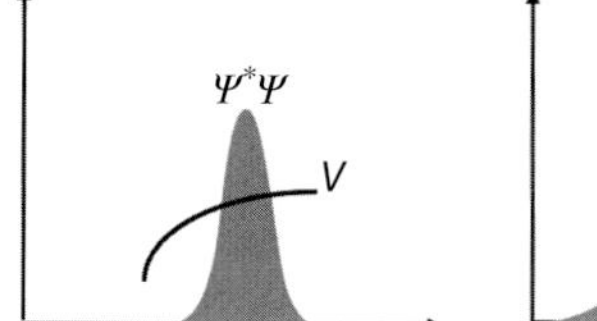

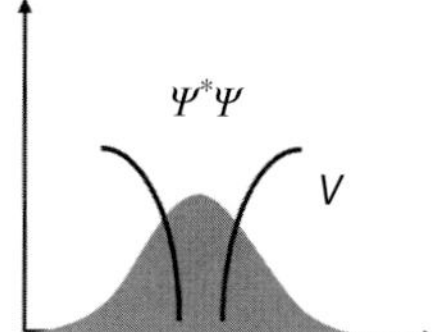

Figure 7.2 Comparison of the potential energy V (solid line) with the localization of the particle (shaded area).

which is analogous to Eq. (7.17). We have, therefore, proved that the classical relation between position and momentum is valid for the quantum mechanical expectation values of these observables.

7.6 The Ehrenfest Theorem

Using Eqs. (7.23) and (7.20), we can also calculate the second derivative of the expectation value for the position, that is, the acceleration:

$$\frac{d^2\langle\hat{x}\rangle}{dt^2} = \frac{d}{dt}\frac{d\langle x\rangle}{dt} = \frac{d}{dt}\frac{\langle\hat{p}_x\rangle}{m} = \frac{1}{m}\frac{d\langle\hat{p}_x\rangle}{dt} = \frac{i}{m\hbar}\langle\hat{H}\hat{p}_x - \hat{p}_x\hat{H}\rangle \tag{7.24}$$

From this it follows[10] that

$$m\frac{d^2\langle\hat{x}\rangle}{dt^2} = \left\langle -\frac{\partial V}{\partial x}\right\rangle = \int \psi^*\left(-\frac{\partial V}{\partial x}\right)\psi\,dx \tag{7.25}$$

This relation between the expectation values is similar to the Newtonian equation, Eq. (5.8). Let us consider two cases as shown in Figure 7.2.

On the right-hand side of the figure, the V potential is changing rapidly in the region where $\psi\psi^*$ is not zero. In that case, the integral in Eq. (7.25) has to be evaluated. In contrast, on the left-hand side, the change of V can be approximated as linear in the region of $\psi\psi^*$. However, if $V \sim x$, then its derivative is constant and can be brought in front of the integral. Since ψ is a normalized function, we get

$$m\frac{d^2\langle\hat{x}\rangle}{dt^2} = \int \psi^*\left(-\frac{\partial V}{\partial x}\right)\psi\,dx \approx \left(-\frac{\partial V}{\partial x}\right)\int \psi^*\psi\,dx = \left(-\frac{\partial V}{\partial x}\right) = F \tag{7.26}$$

which is the Newtonian equation of motion itself, and the wave function plays no role anymore. This line of thought by *Paul Ehrenfest* allows classical mechanics to be regarded as a borderline case of quantum mechanics: *If the potential is*

10 Using Eq. (7.14), we get
$\frac{d^2\langle\hat{x}\rangle}{dt^2} = \frac{i}{m\hbar}\langle\hat{H}\hat{p}_x - \hat{p}_x\hat{H}\rangle = \frac{i}{m\hbar}\left(\frac{1}{2m}\langle\hat{p}^2\hat{p}_x - \hat{p}_x\hat{p}^2\rangle + \langle V\hat{p}_x - \hat{p}_x V\rangle\right) = \frac{i}{m\hbar}\langle V\hat{p}_x - \hat{p}_x V\rangle$ because it makes no difference if we make the first derivation first and the second next, or vice versa. Taking into account the definition of the momentum operator in Eq. (7.5), and using the rule of derivation, Eq. (B.12), we obtain $\frac{i}{m\hbar}(V\hat{p}_x - \hat{p}_x V)\psi = \frac{i}{m\hbar}\left(V\frac{\hbar}{i}\frac{\partial}{\partial x} - \frac{\hbar}{i}\frac{\partial}{\partial x}V\right)\psi = \frac{1}{m}\left(V\frac{\partial\psi}{\partial x} - \frac{\partial V\psi}{\partial x}\right) = -\frac{1}{m}\frac{\partial V}{\partial x}\psi$
Using this result in the expectation value on the right-hand side of Eq. (7.24) follows.

changing slowly in the region where the particle is localized (i.e., where $|\psi|^2 \neq 0$), classical mechanics remains valid. In other words, quantum mechanics does not make classical mechanics invalid, only extends it for rapidly changing potentials in the "size range" of the particle!

Summary in Short

- Heisenberg's commutation rule for the operators of observables, which cannot be accurately measured in the same state simultaneously:
  Postulate 4: **Operators representing the measurement of canonically conjugated observables must fulfill the commutation relation:** $[\hat{p}, \hat{q}] = \hbar/i$.
  From this follows the uncertainty principle $\Delta p \Delta q \geq \hbar/2$.
- Schrödinger's choice for the operators of canonically conjugated observables is:

$$\hat{q} =: q\cdot; \quad \hat{p} =: \frac{\hbar}{i}\frac{\partial}{\partial q}$$

- Classical relations between the observables are retained in quantum mechanics, for example,

$$\hat{\mathbf{L}} = \hat{\mathbf{r}} \times \hat{\mathbf{p}}; \quad \hat{T} = \hat{p}^2/(2m); \quad \hat{E} = \hat{H} = \hat{T} + \hat{V}; \quad \hat{V} = V(\mathbf{r}, t)\cdot$$

- Only the magnitude and one component of the angular momentum can simultaneously be determined.
- The time-dependent Schrödinger equation determines the wave function:
  Postulate 5: **The wave function must fulfill the equation** $-(\hbar/i)(\partial\psi(\mathbf{r},t)/\partial t) = [\hat{T} + \hat{V}]\psi(\mathbf{r}, t)$.
- Time evolution of the observables can be described by $d\langle O\rangle/dt = (i/\hbar)\langle[\hat{H}, \hat{O}]\rangle$. Application to the case of position leads to $v_x = d\langle x\rangle/dt = \langle p_x\rangle/m$.
- *Ehrenfest theorem:* **Classical mechanics is a borderline case of quantum mechanics and is valid if the potential changes slowly in the region where the particle is localized.**

7.7 Questions and Exercises

Problem 7.1 Under what condition can two observables be accurately measured in the same state simultaneously?

Problem 7.2 Are the operators of the various coordinates, x, y, and z commutative? Check it on a function $\varphi(x,y,z)$!

Problem 7.3 Demonstrate that the operators of p_x and y are commutative!

Problem 7.4 Prove that the operator for the absolute square of the angular momentum, and the operator of its z-component are commutative!

Problem 7.5 What is the operator for the x-component of the angular momentum?

Problem 7.6 What consequence does the uncertainty principle have for the Bohr orbitals?

Problem 7.7 What is the current density delivered by a single electron, provided that its wave function is $\psi(x,t) = (1/\sqrt{L})e^{i(kx-\omega t)}$?

8

Quantum Mechanical States

In this chapter...
The five postulates we have found so far suffice to describe a *single particle*:

- Postulate 1 describes our concept of a particle by restricting the possible wave functions.
- Postulate 5 allows us to determine the wave function with knowledge of the potential energy. However, care should be taken to select the solution according to Postulate 1 from all the mathematically possible ones.
- Postulates 2–4 describe the way how the outcome of the measurement of a classically defined observable can be predicted. From these postulates, it follows that in the case of a mixed state, only the expectation (i.e., mean) value of many measurements and the uncertainty (standard deviation) with respect to that can be predicted. For this purpose, the operators of Schrödinger and the wave function (determined by using Postulates 1 and 5) should be used.

If the initial state was a mixed state, the measurement forces the particle into one of the eigenstates, resulting in the corresponding eigenvalue. If the state to be measured was an eigenfunction of the operator to begin with, the expectation value will be the corresponding eigenvalue and the uncertainty will be zero. Therefore, it is very important to know the eigenstates and eigenvalues of the classically defined observables in advance. These can be obtained by solving the eigenvalue equation of the corresponding operator. Here, we are going to investigate the eigenvalue equation for position x and momentum p_x first, then turn to the eigenvalue equation of the energy in the one-dimensional case ($V = V(x)$). We come back to the three-dimensional eigenvalue equations of the angular momentum and of the energy, in the case of a central field $V(r)$, in Chapter 11.

8.1 Eigenstates of Position

Considering the definition of the position operator in Eq. (7.5), the eigenvalue equation for the position reads as

$$\hat{x} =: x\cdot \; \Rightarrow x\cdot\varphi_i(x) = x_i\varphi_i(x) \tag{8.1}$$

Essential Quantum Mechanics for Electrical Engineers, First Edition. Peter Deák.

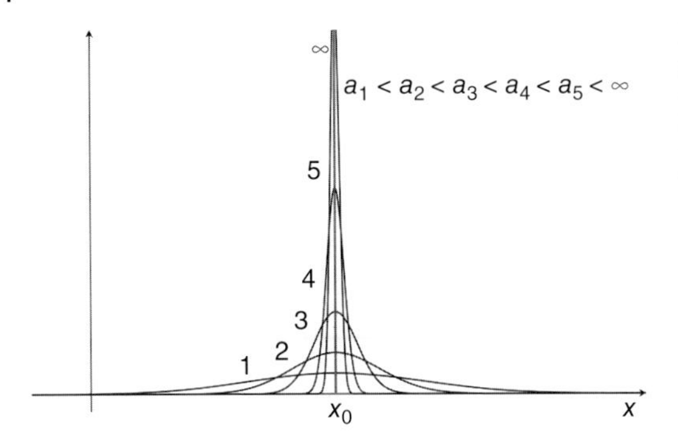

Figure 8.1 The generalized function of Eq. (8.2), displayed for increasing integration limits *a*. The Dirac delta function corresponds to $a = \infty$. Color online.

where the eigenvalue x_i is one possible value of the position and $\varphi_i(x)$ is the corresponding eigenstate. This equation requires that the multiplication of $\varphi_i(x)$ by its variable (i.e., by the value x at any position x) should lead to the same result as the multiplication by a constant (x_i at the given position). At first sight, it appears to be impossible to find a function for which this requirement could be satisfied. Indeed, a solution cannot be found among the ordinary functions. However, *Paul A. M. Dirac*, the first to consider this problem, has shown that the following integral fulfills the requirement:

$$\delta(x - x_0) = \lim_{a \to \infty} \frac{1}{2\pi} \int_{-a}^{a} e^{ik(x-x_0)}\, dk \tag{8.2}$$

This function, called the Dirac delta ever since, is a so-called *generalized function*, and – as can be seen in Figure 8.1 – is zero at any x, except for x_0 where it is infinite.

A function $\varphi_i = \delta(x - x_i)$ will have zero value at any $x \neq x_i$ position and ∞ at x_i. This will also remain so even after multiplication by either x or x_i. Since the Dirac delta function is not changed by the position operator, it is an eigenfunction of the position

$$x \cdot \delta(x - x_i) = x_i \delta(x - x_i) \tag{8.3}$$

Since one can construct a Dirac delta at any conceivable point on the x-axis or, in three dimensions, at any point in space, every point is an allowed eigenvalue of position (continuous eigenvalue spectrum).

The $\delta(x - x_i)$ eigenfunctions of position are single valued and twice derivable.[1] It can be proven that the Dirac delta has the following property for any function $f(x)$:

$$\int_{-\infty}^{+\infty} f(x) \cdot \delta(x - x_0) \cdot dx = f(x_0) \tag{8.4}$$

From this follows that the Dirac delta cannot be square-integrable.[2] It is not continuous either, that is, it is not a regular function. *According to Postulate 1,*

---

1 Because each component in the summation, expressed by the integral of Eq. (8.1), is twice derivable.

2 Using Eq. (8.4), it follows that the scalar product of the Dirac delta with itself is $\int_{-\infty}^{+\infty} \delta(x - x_0) \cdot \delta(x - x_0) \cdot dx = \delta(0) = \infty$.

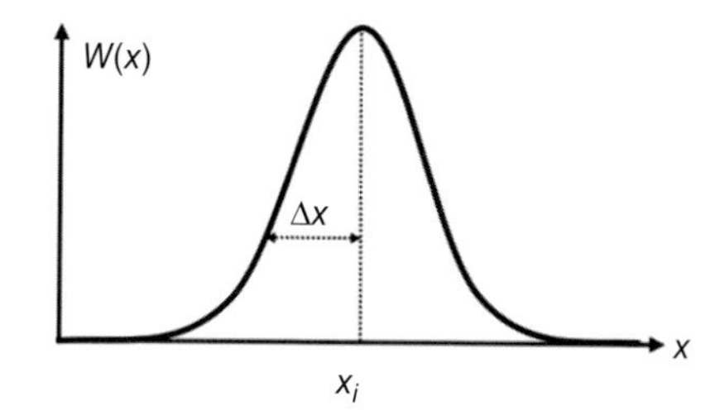

Figure 8.2 Probability density for measuring the position x_i.

however, this means that a particle can never be in an eigenstate of position. The particle is simply not a point mass and its position can only be measured with some uncertainty.

N.B.: Postulate 4 only excludes the accurate measurement of position and momentum simultaneously (on the same state) but would, in principle, allow $\Delta x = 0$ at the cost of $\Delta p_x = \infty$. However, the regularity condition of Postulate 1 forbids a state for the particle in which it would be localized to a geometrical point.

From the viewpoint of measuring the position, a particle is, therefore, always in a mixed state. According to Postulate 3, we can only calculate the probability with which any x_i point in space could be measured as position of the particle. Using Eqs (6.11) and (8.4),

$$W(x_i) = |\langle\psi|\varphi_i\rangle|^2 = \left|\int \psi^*(x,t)\delta(x-x_i)dx\right|^2 = |\psi(x_i,t)|^2 \tag{8.5}$$

As we can see, Postulate 3 reproduces the Born–Jordan interpretation of the absolute square of the wave function as probability density. In the double-slit experiment, we have attempted to detect the position of the electron. We have obtained spots on the CCD screen, and the intensity distribution across one spot is shown in Figure 8.2. This corresponds to a finite wave packet, that is, the measurement has localized the particle.

8.2 Eigenstates of Momentum

Considering the definition of the momentum operator in Eq. (7.5), the eigenvalue equation for the position operator is

$$\hat{p}_x =: \frac{\hbar}{i}\frac{\partial}{\partial x} \Rightarrow \frac{\hbar}{i}\frac{\partial}{\partial x}\cdot\varphi_n(x) = p_{xn}\varphi_n(x) \tag{8.6}$$

where p_{xn} is a possible eigenvalue of the momentum and $\varphi_n(x)$ is the corresponding eigenstate. The eigenfunctions of the differential operator are, in analogy to Eq. (B.21), the exponential functions $\varphi_n(x) = \exp(ip_{xn}x/\hat{h})$, because

$$\frac{\hbar}{i}\frac{\partial}{\partial x}e^{i\frac{p_n}{\hbar}x} = p_{xn}e^{i\frac{p_n}{\hbar}x} \tag{8.7}$$

with arbitrary p_{xn}. Just as the position, the momentum can also vary continuously. However, similar to the case of the position, such momentum eigenstates are not regular, because they are not square-integrable over the whole space either:

$$\int_{-\infty}^{+\infty} \left(e^{\frac{i}{\hbar}p_{xn}x}\right)^* e^{\frac{i}{\hbar}p_{xn}x}\, dx = \int_{-\infty}^{+\infty} e^{-\frac{i}{\hbar}p_{xn}x} e^{\frac{i}{\hbar}p_{xn}x}\, dx = \int_{-\infty}^{+\infty} 1\, dx = \infty \tag{8.8}$$

Therefore, according to Postulate 1, a particle cannot be in a momentum eigenstate if there is no potential restricting its motion. In other words, one cannot measure an accurate value of the momentum on a freely moving particle! The condition of regularity forbids not only completely localized but also completely delocalized states. This means that both basic concepts of classical mechanics, point mass and infinite wave, are impermissible idealizations.

N.B.: Upon comparison with Eq. (A.18), one can see that the eigenfunctions of the momentum operator constitute the position-dependent part ($\exp(ikx)$) of a harmonic wave. This means that a harmonic wave, $\exp[-i(\omega t - k_{xn}x)]$, is also an eigenfunction of the momentum operator.

$$\frac{\hbar}{i}\frac{\partial}{\partial x} e^{-i(\omega t - k_{xn}x)} = \hbar k e^{-i(\omega t - k_{xn}x)} \tag{8.9}$$

Comparison of Eqs (8.9) and (8.7) provides the hypothesis of de Broglie (Eq. (4.1)) automatically:

$$k_{xn} = \frac{p_{xn}}{\hbar} \tag{8.10}$$

If a potential is restricting the motion of the particle to a range of length L (as, e.g., in Section 4.5 to $L = 2r\pi$), *the momentum eigenfunctions can be normalized to L:*

$$\int_0^L \frac{1}{\sqrt{L}} e^{-ik_{xn}x} \frac{1}{\sqrt{L}} e^{ik_{xn}x}\, dx = \frac{1}{L}\int_0^L 1\, dx = 1$$

So we obtain an allowed (regular) momentum eigenstate

$$\varphi_n(x) = \frac{1}{\sqrt{L}} e^{\frac{1}{\hbar}p_{xn}x} \tag{8.11}$$

However, in such cases, *the momentum eigenvalues will be quantized* (details are provided in Chapter 11):

$$p_{xn} = \hbar k_{xn} = \frac{2\pi n\hbar}{L} \tag{8.12}$$

If, following de Broglie, we assume the electron to move along a circle with $L = 2r\pi$, we obtain for the allowed momentum eigenvalues the same result as in Eq. (4.13).

8.3 Eigenstates of Energy – Stationary States

We have seen so far that the assumptions made in the interpretation of experiments on the electron can be derived from the postulates. The quantization of the electron energy in the hydrogen atom should follow from the energy eigenvalue equation with $V(r) = -[1/(4\pi\varepsilon_0)](e^2/r)$, but that is a three-dimensional problem, which is mathematically not easy to solve (see Chapter 11). Here, we consider first the general problem of a *time-independent potential energy* in the time-dependent Schrödinger equation (Postulate 5):

$$-\frac{\hbar}{i}\frac{\partial\psi(\mathbf{r},t)}{\partial t} = -\frac{\hbar^2}{2m}\Delta\psi(\mathbf{r},t) + V(\mathbf{r})\cdot\psi(\mathbf{r},t) \tag{8.13}$$

This equation contains operators that act on either the time or the space variable only. In such cases, one can assume that the solution can be written as the product of a solely time- and a solely position-dependent function:

$$\psi(\mathbf{r},t) = \varphi(\mathbf{r})\cdot\tau(t) \tag{8.14}$$

Substituting this into Eq. (8.13) gives

$$-\frac{\hbar}{i}\varphi(\mathbf{r})\frac{d\tau(t)}{dt} = \tau(t)\left[-\frac{\hbar^2}{2m}\Delta\varphi(\mathbf{r}) + V(\mathbf{r})\varphi(\mathbf{r})\right] \tag{8.15}$$

which can be rearranged in such a way that all time-dependent terms are on the left-hand side and all position-dependent terms are on the right-hand side:

$$-\frac{\hbar}{i}\frac{1}{\tau}\frac{d\tau}{dt} = \frac{1}{\varphi}\left[-\frac{\hbar^2}{2m}\Delta\varphi + V(\mathbf{r})\varphi\right] \tag{8.16}$$

This equation implies that two functions of different variables should always be equal, independent of the actual values of their variables. This is only possible if both functions are constant. Let us call the constant E. This yields two equations:

$$-\frac{\hbar}{i}\frac{1}{\tau}\frac{d\tau}{dt} = E \rightarrow -\frac{\hbar}{i}\frac{d\tau_n}{dt} = E_n\tau_n \tag{8.17}$$

$$\frac{1}{\varphi}\left[-\frac{\hbar^2}{2m}\Delta\varphi + V(\mathbf{r})\varphi\right] = E \rightarrow \left[-\frac{\hbar^2}{2m}\Delta\varphi_n + V(\mathbf{r})\varphi_n\right] = E_n\varphi_n \tag{8.18}$$

As can be seen, the rearrangement leads to the eigenvalue equation of the two possible energy operators in Eqs (7.11) and (7.14).[3] From Eq. (8.17), it follows that *in case of a time-independent potential the time-dependent part of the wave function is always*

$$\tau_n(t) = e^{-i\frac{E_n}{\hbar}t} \tag{8.19}$$

Equation (8.18) is the eigenvalue equation of the Hamilton operator,

$$\left[-\frac{\hbar^2}{2m}\Delta\varphi_n + V(\mathbf{r})\varphi_n\right] = E_n\varphi_n \tag{8.20}$$

which provides the position-dependent part of the wave function. The wave function is, therefore,

$$\psi_n(\mathbf{r},t) = \varphi_n(\mathbf{r})e^{-i\frac{E_n}{\hbar}t} \tag{8.21}$$

Since this function is a solution for both Eqs (8.20) and (8.13), the energy eigenvalue equation (8.20) is also sometimes described as the *time-independent*

3 That is why it made sense to choose the constant in Eqs (8.17) and (8.18) as E. The indices n enumerate the possible eigenfunctions.

Schrödinger equation. Our result also implies that in a time-independent potential energy $V(\mathbf{r})$, that is, in a conservative force field, the wave function is an energy eigenstate and neither the state nor the energy changes in time. In other words, the law of energy conservation remains valid also in quantum mechanics.

N.B.: The fact that the energy eigenstates are stationary also follows from the uncertainty relation of Eq. (7.3) for time and energy

$$\Delta t \Delta E \geq \frac{\hbar}{2} \tag{8.22}$$

In an energy eigenstate $\Delta E = 0$, the lifetime must, therefore, be $\Delta t = \infty$.[4]

8.4 Free Motion

As shown in the previous subsection, stationary states of a particle are energy eigenstates. Therefore, an important question is: what other observables can characterize an energy eigenstate? The answer to this question can be found with the help of the von Neumann theorem (Section 6.4): those observables that have an operator commutative with the Hamilton operator.[5] Obviously, stationary energy eigenstates cannot be, at the same time, eigenfunctions of position, because the operator $\mathbf{r}\cdot$ is not commutative with the operator of the kinetic energy which, as defined in Eq. (7.12), contains the Laplace operator (second derivative).

A stationary state can only be an eigenstate of momentum if the Hamilton operator does not contain a $V(\mathbf{r})$ function of the coordinates, since that would not be commutative with the momentum operator which, according to Eq. (7.5), contains the nabla (∇) operator (first derivative). Since the zero point of the potential energy scale can be freely chosen, it can be set for any $V(\mathbf{r}) = const.$ potential energy at the value of the constant. So, if the potential energy does not depend on either time or position, the Hamilton operator of Eq. (8.20) contains only the kinetic energy operator of Eq. (7.13), and

$$\hat{T}\varphi_n(\mathbf{r}) = \frac{\hbar^2}{2m}\Delta\varphi_n(\mathbf{r}) = E_n\varphi_n(\mathbf{r}) \tag{8.23}$$

The second derivative Δ in this operator is commutative with the first derivative ∇ of the momentum operator, so the eigenstates of the momentum in Eq. (8.11) are also eigenstates of the kinetic energy. According to Eq. (8.21), the corresponding wave function in a one-dimensional case is then

$$\psi_n(x,t) = \varphi_n(x)e^{i\frac{E_n}{\hbar}t} = \frac{1}{\sqrt{L}}e^{\frac{i}{\hbar}p_{xn}x}e^{i\frac{E_n}{\hbar}t} = \frac{1}{\sqrt{L}}e^{\frac{i}{\hbar}(p_{xn}x - E_n t)} \tag{8.24}$$

---

4 Spontaneous emission, by which excited electrons return to the ground state (see Section 3.5), seems to contradict the infinite lifetime of energy eigenstates. One should consider, however, that an uncertainty principle must apply to the E-field as well, excluding the possibility of the field being at complete rest. EM waves arising from fluctuations (so-called zero Kelvin vibrations) of the field stimulate the "spontaneous" emission.

5 N.B.: This can also be seen from Eq. (7.19). If $[\hat{H}, \hat{O}] = 0$, the expectation value of the observable does not change in time.

This is a harmonic wave

$$\psi_n(x,t) = \frac{1}{\sqrt{L}} \exp[i(k_n x - \omega_n t)] \tag{8.25}$$

with wave number $k_{xn} = p_{xn}/\hbar$ and angular frequency $\omega_n = E_n/\hbar$, just as it was assumed by de Broglie in Eq. (4.11) for an electron-wave propagating on a circle of circumference L. Substituting Eq. (8.25) into Eq. (8.23), we obtain

$$\hbar\omega_n = E_n = \frac{p_{xn}^2}{2m} = \frac{\hbar^2 k_{xn}^2}{2m} \tag{8.26}$$

This is the *dispersion relation of the free electron.* As we have guessed in Eqs (5.1) and (5.2), ω_n is a nonlinear function of k.

N.B.: It might seem contradictory at the first sight that we have assumed $V(x) = 0$, while a potential must confine the electron wave to a circular orbit of length L. However, the x coordinate is meant here as the position along the circle, that is, $x = R\phi$, and no force is acting on the electron in the tangential direction. It is the centripetal force (i.e., a $V(r)$ potential), which keeps the electron on orbit. In Chapter 11, we see that this model of de Broglie is only a rough approximation to the real quantum mechanical problem of an electron in the hydrogen atom.

If the particle could really move freely in the entire space (in other words, if $V(\mathbf{r})$ was really zero everywhere), the simultaneous eigenfunction of $\hat{T}$ and $\hat{p}$ would be an infinite harmonic wave. According to Postulate 1, a particle cannot be in such a state. However, superposition of harmonic waves with wave numbers in the range $[k_0 - \infty, k_0 + \infty]$ and with a Gaussian distribution of amplitudes

$$\psi_0(k) = \frac{d}{\sqrt{\pi}} e^{-d^2(k-k_0)^2}$$

(where d is the width of the distribution) results in a normalized wave packet,

$$\psi(x,t) = \left[\frac{e^{\frac{(v_g t - x)^2}{4d^2(1-iBt/d^2)}}}{(1 - iBt/d^2)^{1/2}} \right] e^{i(\omega_0 t - k_0 x)t} \tag{8.27}$$

Here ω_0, v_g, and B are the coefficients of the first three terms in the Taylor expansion of $\omega(k)$ around k_0. Since this normalized wave packet is square-integrable and is a sum of harmonic waves (which are single-valued, continuous, and twice derivable functions), it is compatible with the requirements of Postulate 1; that is, it is an allowed state for the particle. It also satisfies the time-dependent Schrödinger equation of the free motion:

$$-\frac{\hbar}{i}\frac{\partial\psi(x,t)}{\partial t} = -\frac{\hbar^2}{2m}\frac{\partial^2\psi(x,t)}{\partial x^2} \tag{8.28}$$

Equation (8.28) is a wave equation (see also Section 5.7), satisfied by any harmonic wave. Therefore, the sum of harmonic waves, that is, the wave packet, is also a solution.

However, due to the nonlinear dispersion relation in Eq. (8.26), the phase velocity differs from the group velocity (see Section A.5), and the packet delocalizes with time as shown in Figure 8.3.

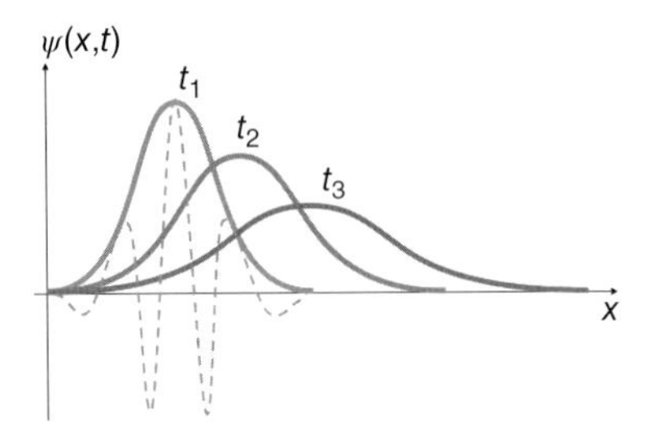

Figure 8.3 Delocalization of the wave packet describing a freely moving electron. Color online.

The state of a freely moving electron is not stationary. This follows also from the fact that the harmonic waves are energy eigenstates, and their sum, the wave packet, is then a mixed state from the viewpoint of the energy. If $\Delta E \neq 0$, then Δt must be finite.

N.B.: An electron gun is a cathode ray tube with a hole in the anode, through which the electrons can leave. A U voltage between the cathode and the anode accelerates the electrons of charge e to an energy $E = eU$, but with some ΔE uncertainty because the accuracy of setting the voltage is limited. Therefore, the state of the emitted electron is, in principle, a wave packet, but very near to being an energy eigenstate, that is, quite delocalized. That is why it can realize both slits in the double-slit experiment. It is the detector that localizes the electron into a very localized wave packet with a large energy uncertainty ΔE. Without the detector, the electron would delocalize even more. We should consider, however, that the world is not empty, and the potential energy cannot be constant in the entire space.[6]

8.5 Bound States

There is, of course, an infinite number of possibilities where the $V(\mathbf{r})$ potential energy is not constant. General observations can be derived for those ranges of space, where $V(\mathbf{r})$ is smaller than the total energy E, while it is bigger outside the region. Let us consider such a case in one dimension, as shown in Figure 8.4. Outside the $[x_1, x_2]$ region $V(x)$ is $> E$, which would mean a kinetic energy of $T = E - V(x) < 0$. According to the classical definition of the kinetic energy (see Eq. (A.8)), this is not possible (momentum would become an imaginary number); therefore, a classical (point-mass-like) particle would be bound to move only between x_1 and x_2. The time-independent Schrödinger equation (eigenvalue equation of the energy) with this $V(x)$ potential is

$$-\frac{\hbar^2}{2m}\frac{d^2}{dx^2}\varphi(x) + V(x)\varphi(x) = E\varphi(x)$$

which can be brought to the standard form of an ordinary differential equation of second order:

$$\frac{d^2\varphi(x)}{dx^2} = -\frac{2m}{\hbar^2}[E - V(x)]\varphi(x) \equiv -\alpha(E,x)\varphi(x) \tag{8.29}$$

6 By the way, just like electrons, photons can only be wave packets. This is the reason why the band width and coherence length of even lasers are finite.

Figure 8.4 Impermissible (E) and allowed (E_n) bound states. Color online.

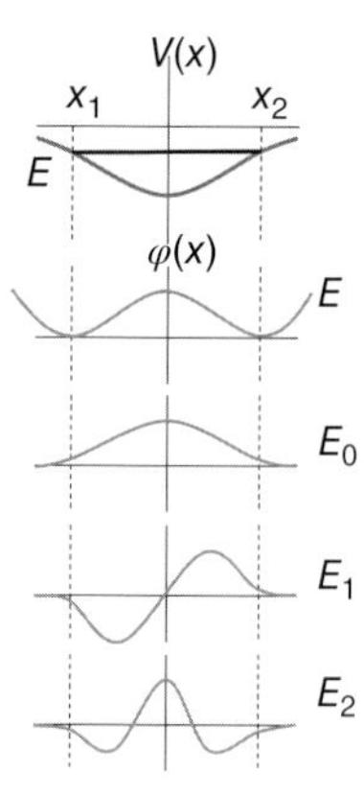

For an arbitrary value of E, let us assume the boundary condition $\varphi(0) > 0$, and consider that

$$\begin{aligned} &\alpha > 0, \quad \text{if } x_1 < x < x_2 \\ &\alpha < 0, \quad \text{if } x_1 > x > x_2 \end{aligned} \tag{8.30}$$

The second derivative determines the curvature of the function. Since in the vicinity of $x = 0$ both $\varphi(x)$ and α are positive, from Eqs (8.29) and (8.30), it follows that the curvature must be negative in both directions. In other words, $\varphi(x)$ must have a maximum at $x = 0$. Outside the $[x_1, x_2]$ region $\alpha(E, x)$ is negative, so the curvature of the function becomes positive. For an arbitrary value of E this means that $\varphi(x)$ goes to infinity when $x \to \pm\infty$, as shown by the first $\varphi(x)$ curve in Figure 8.4. Such states are forbidden by Postulate 1 because they are not square-integrable. However, for some special, discrete E_n values it is conceivable to have a $\varphi(x)$ function with positive curvature at x_1 and x_2 but going to zero for $x \to \pm\infty$. Such a solution of Eq. (8.29) is shown by the second $\varphi(x)$ curve of Figure 8.4. The third curve shows a square-integrable solution for the boundary condition $\varphi(0) = 0$. The fourth curve is also regular and satisfies Eq. (8.29).

This consideration shows that the regularity condition prescribes discrete, *quantized* energy values in a bound state. We have already seen something similar to that: restricting the motion to a finite length has led to quantized momenta in Eq. (8.12), which – according to the dispersion relation in Eq. (8.26) – corresponds to discrete energies. Postulate 1 requires the wave function to vanish at infinity, and this leads to discrete $\nu = E/\hbar$ frequencies just as in the case of classical waves of a string with fixed ends.

The solutions of the Schrödinger equation with potential energy similar to that in Figure 8.4 are called *bound states*. Still, the particle is not entirely restricted to the $[x_1, x_2]$ region! For finite $V(x)$ values, the $\varphi(x)$ curves go only slowly to zero outside $[x_1, x_2]$, as can be seen in Figure 8.4. This means that the probability density to find the particle, $|\varphi(x)|^2$, does not become zero at once outside $[x_1, x_2]$ either. The quantum mechanical particle can also be found in the region forbidden for a classical one! (This is the reason for the tunnel effect, see Chapter 10.)

So we can see that the energies of the bound states must be quantized. The states with energies E_0, E_1, and E_2 in Figure 8.4 all belong to the same $V(x)$ potential energy, so they differ only in their kinetic energy. Using Eqs (8.23) and (6.12), the expectation value of the kinetic energy is

$$\langle T\rangle = \langle\varphi|\hat{T}\varphi\rangle = \left\langle\varphi\left|-\frac{\hbar^2}{2m}\frac{d^2}{dx^2}\varphi\right.\right\rangle \tag{8.31}$$

which, for a given φ, is determined by its second derivative, that is, by the curvature of the function. Since with increasing number of nodes in the same range, the curvature must increase, the energies of the depicted states have the order $E_0 < E_1 < E_2$. (In other words, the higher the "frequency" in Figure 8.4, the higher is the energy.) It is generally true that bound states can be energetically ordered according to the number of nodes in the wave function, and the ground state E_0 has no nodes.

Summary in Short

- The eigenstates of position in space are the Dirac delta functions, $fi_i = \delta(\mathbf{r} - \mathbf{r}_i)$, with any possible $\mathbf{r} = \mathbf{r}_i$ as eigenvalue. Such states are, however, not regular functions and are not permitted for a particle. From the viewpoint of measuring the position, every possible ψ state of a particle is a mixed state. The probability of finding the particle in a $d\mathbf{r}$ range around the point $\mathbf{r}_i$ is $W(\mathbf{r}_i) = |\langle\psi|\varphi_i\rangle|^2 d\mathbf{r} = |\psi(\mathbf{r}_i, t)|^2 d\mathbf{r}$.
- The eigenstates of momentum are the $e^{i\mathbf{kr}}$ functions with arbitrary $\mathbf{p} = \hat{h}\mathbf{k}$ eigenvalues. Such states are, however, not regular if extending infinitely. A particle can only be in a momentum eigenstate if its motion is confined to a finite volume by a potential. In this case, however, only discrete, quantized momentum values are allowed. Those can be measured without uncertainty.
- If the potential energy does not depend on time, the wave function is determined by the eigenfunctions and eigenvalues of the time-independent Schrödinger equation, that is, $[\hat{T} + \hat{V}(\mathbf{r})]\varphi_n(\mathbf{r}) = E_n\varphi_n(\mathbf{r})$, as $\psi(\mathbf{r}, t) = \varphi_n(\mathbf{r})e^{i\frac{E_n}{\hbar}t}$.
- A free-moving particle can be described by a wave packet. From the viewpoint of the energy, this is a mixed state and, due to dispersive propagation, it delocalizes with time. A particle moving freely in restricted space can have a state corresponding to a harmonic wave, which is a simultaneous eigenfunction of momentum and kinetic energy. The dispersion relation for such free electrons is $E_n(k) = \frac{\hbar^2 k_n^2}{2m}$.
- The regularity condition of Postulate 1 allows only bound states that go to zero at infinity. Such wave functions belong to certain discrete (quantized) energies. The (kinetic) energy increases with the number of nodes in the wave function. For $E - V = T < 0$, the wave function does not become at once zero, so the probability to find the particle outside the classically allowed region decays only slowly.
- The way of solving quantum mechanical problems is shown in Figure 8.5.

8.6 Questions and Exercises

Problem 8.1 A particle moving in the potential $V(x) = (1/2)Dx^2 = (1/2)m\omega_0^2 x^2$ is called a quantum mechanical oscillator because $x = x_0 \sin(\omega_0 t)$

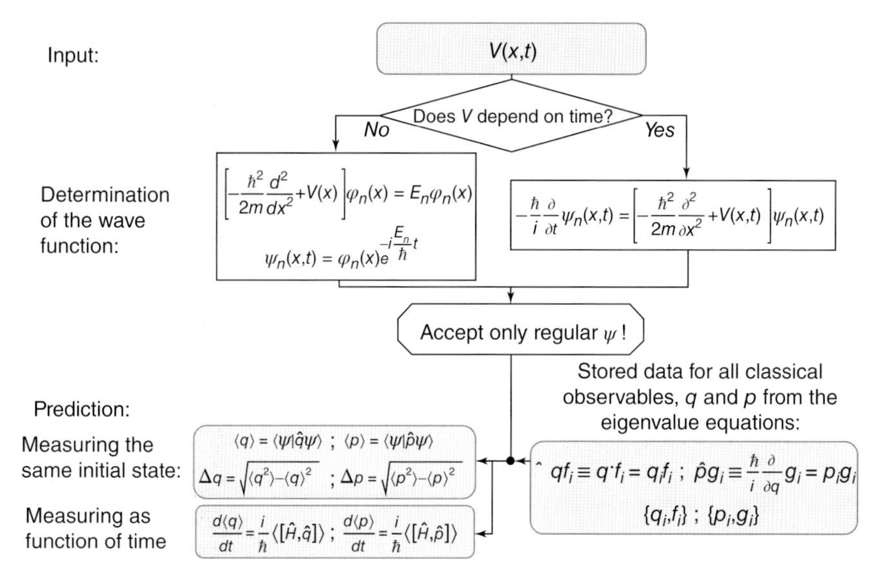

Figure 8.5 Flow chart of solving problems in quantum mechanics.

still holds (x_0 is the amplitude of the vibration). Derive a relation between the uncertainty of position, Δx, and the uncertainty of momentum, Δp_x, for this case, using Eq. (7.21) and the trivial trigonometric relations $\langle \sin(\omega_0 t)\rangle = \langle \cos(\omega_0 t)\rangle$ and $\langle \sin^2(\omega_0 t)\rangle = \langle \cos^2(\omega_0 t)\rangle$!

Problem 8.2 Using the previous result, determine the dependence of both uncertainties on the properties of the oscillator (m and ω_0), using Eq. (7.3)!

Problem 8.3 In contrast to the classical harmonic oscillator, the energy of such bound states is quantized. Based on the uncertainties derived in the previous problem, estimate the lowest allowed energy! Why must this be larger than zero?

Problem 8.4 Atoms of a solid can be considered to be harmonic oscillators. Using Planck's hypothesis for the energy absorption of a metal wall, estimate the quantum mechanically allowed energies of the atomic vibrations, taking also into account the minimal energy obtained here! If you did everything right, your result will agree with the energy eigenstates obtained by solving the time-independent Schrödinger equation of a quantum mechanical oscillator exactly.

Problem 8.5 The solution of the time-independent Schrödinger equation for the harmonic oscillator gives the following ground state:

$$\varphi_0 = \frac{1}{(\pi x_0^2)^{1/4}} e^{-\frac{1}{2}\left(\frac{x}{x_0}\right)^2}$$

Calculate the probability density of the particle at $x = 2x_0$ and interpret the result from the viewpoint of classical mechanics!

9

The Quantum Well: the Basis of Modern Light-Emitting Diodes (LEDs)

In this chapter...
We consider a practical application of quantum mechanics. In contrast to most other cases, the time-independent Schrödinger equation, to be solved here, contains only a simple, one-dimensional $V(x)$ potential, and the solution can be found without spending much computer time. In Chapters 2–4, we have seen how studying the physics of light bulbs and discharge lamps has led to the introduction of quantum mechanics and quantum optics (and eventually to the discovery of the laser). In this chapter, we see how the conscious application of this new physics leads to the most recent stage of lighting technology, the quantum-well light-emitting diodes (LEDs).

9.1 Quantum-Well LEDs

We have shown in Section 8.5 that the electron can have only quantized energies in a bound state. In Section 4.3, we have also seen that the quantized energy values (also called energy levels) in the hydrogen atom form a series according to $\sim 1/n^2$. However, a solid contains about 10^{23} atoms with many atomic energy levels lying close to each other and forming the so-called energy band,[1] as we have already seen in Figure 3.2 when discussing the photoelectric effect. While explaining the photovoltaic effect, it has already been mentioned that in semiconductors there is a fully occupied and a fully empty band, and current can only flow if electrons are promoted from the former, the so-called valence band, into the latter, the so-called conduction band, by, for example, photon absorption. The reverse process is, however, also conceivable.

As shown in Figure 9.1, an electron in the conduction band can "fall" into a hole (unoccupied level) in the valence band, and emit the energy difference between the two levels as a photon. If electric currents provide a supply of electrons in the conduction band and holes in the valence band, we will have an almost monochromatic light source with a frequency corresponding to the

1 The reasons for this can only be explained in the planned book *Essential Semiconductor Physics for Electrical Engineers.*

Essential Quantum Mechanics for Electrical Engineers, First Edition. Peter Deák.

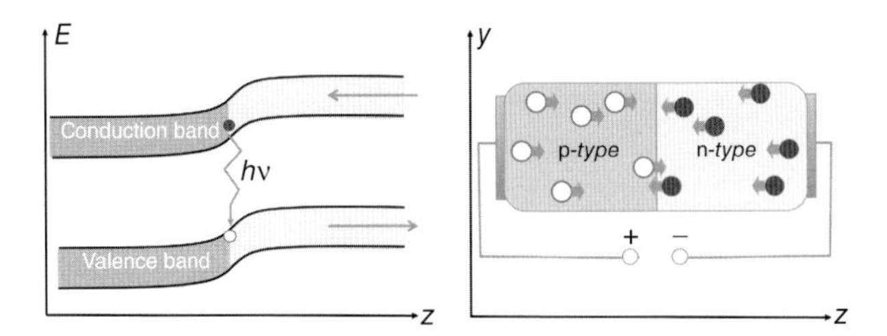

Figure 9.1 Schematic representation of a simple light-emitting diode. Electrons (blue) in the conduction band are negative, the holes (red) in the valence band are positive charge carriers. Color online.

energy difference between the band edges. Such a supply is available at the so-called n–p junction of an LED. Such primitive LEDs have, however, two disadvantages. First, the frequency (color) of the LED is completely prescribed by the chosen semiconductor. Second, and more importantly, electrons and holes will meet in space only rarely, and the efficiency will be very low.

The position of the band edges depends on the potential energy $V(z)$, which is characteristic to the given material. In 1971, *Leo Esaki* suggested to vary the composition of a compound semiconductor along the z-axis by applying, for example, a $Ga_{1-x}Al_xAs/GaAS/Ga_{1-x}Al_xAs$ sandwich structure. Such layers can be grown on top of each other and, provided that $x < 0.4$, the crystal structure remains the same. The composition parameter x can be changed quite quickly so, to a good approximation, the change of the potential between $Ga_{1-x}Al_xAs$ and GaAs layers can be regarded as abrupt. Since the potential of the holes is approximately a mirror image of that of the electrons, the potential energy changes for electrons and holes, as depicted in Figure 9.2. We are going to see in the following that such a potential results in well-defined, discrete energies both for the electrons and the holes and that their energy difference, and with that the emitted photon frequency, can be tuned by the Al-fraction x and by the width of the GaAs part.

In quantum mechanics, such a potential is called a quantum well. Actually, in our case, quantum ditch would be a better name. Electrons coming from the $Ga_{1-x}Al_xAs$ part will fall into the "ditch" and so will the holes.[2] Electrons and holes are, therefore, both trapped in a thin layer and can recombine with a high probability. Similar to the idea by Esaki, the realization by *Zhores Alferov* has been acknowledged by a Nobel Prize in physics, in 1973 and 2000, respectively. The idea of the quantum well (QW)-LED is used in all semiconductor lasers and, ever since the realization of the blue (and UV) LEDs, for which *Shuji Nakamura* received the Nobel Prize in 2014, it is also being applied in lighting.

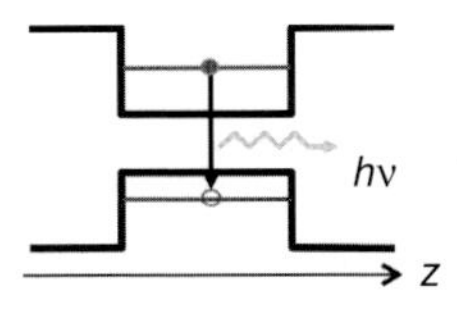

Figure 9.2 Electron and hole potentials (upper and lower thick line, respectively) in a $Ga_{1-x}Al_xAs/GaAs/Ga_{1-x}Al_xAs$ semiconductor heterostructure sandwiched in the z direction.

2 Again, the reason for the holes "falling" upward can only be given in the planned book *Essential Semiconductor Physics for Engineers*.

9.2 Energy Eigenvalues in a Finite Quantum Well

In order to calculate the energies of the electrons and holes bound in the QW (and to design the color of the LED), the time-independent Schrödinger equation, Eq. (8.20), has to be solved in one dimension for the potential energy shown in Figure 9.3,

$$V(x) = \begin{cases} 0, & \text{if } |x| < a \\ V_0, & \text{if } |x| \geq a \end{cases} \tag{9.1}$$

Since $V(x)$ is not a continuous function, we have to construct the Schrödinger equation in three regions separately, and then merge the three solutions at the boundaries in such a way that – as Postulate 1 requires – they form a continuous, twice derivable, and square-integrable function. The Schrödinger equations in the three regions are as follows:

$$x < -a: \quad -\frac{\hbar^2}{2m}\frac{d^2\varphi_1}{dx^2} + V_0\varphi_1 = E\varphi_1 \rightarrow \frac{d^2\varphi_1}{dx^2} = \alpha^2(E)\varphi_1 \tag{9.2}$$

$$-a < x < a: \quad -\frac{\hbar^2}{2m}\frac{d^2\varphi_2}{dx^2} = E\varphi_2 \rightarrow \frac{d^2\varphi_2}{dx^2} = -\beta^2(E)\varphi_2 \tag{9.3}$$

$$x > a: \quad -\frac{\hbar^2}{2m}\frac{d^2\varphi_3}{dx^2} + V_0\varphi_3 = E\varphi_3 \rightarrow \frac{d^2\varphi_3}{dx^2} = \alpha^2(E)\varphi_3 \tag{9.4}$$

where we have introduced the functions:

$$\alpha^2(E) \equiv \frac{2m}{\hbar^2}(V_0 - E); \quad \beta^2(E) \equiv \frac{2m}{\hbar^2}E \tag{9.5}$$

(the square emphasizing that they are always positive). The differential equations in Eqs (9.2)–(9.4) are satisfied by the following functions:

$$\varphi_1(-a > x) = Ae^{\alpha x} + Be^{-\alpha x} \tag{9.6}$$

$$\varphi_2(-a < x < a) = C\sin\beta x + D\cos\beta x \tag{9.7}$$

$$\varphi_3(x > a) = Ee^{\alpha x} + Fe^{-\alpha x} \tag{9.8}$$

where we have considered the mathematical theorem that the general solution of a differential equation can be constructed as the linear combination of any two special solutions. The functions $e^{\pm\alpha x}$ satisfy Eqs (9.2) and (9.4), both $\sin(\beta x)$ and

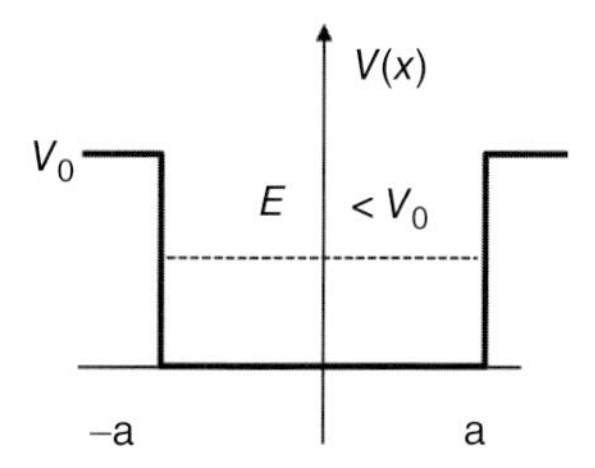

Figure 9.3 The one-dimensional potential well.

$\cos(\beta x)$ satisfy Eq. (9.3), and A, B, C, D, E, and F are parameters to be determined from the regularity condition of Postulate 1.

To satisfy the requirement of square integrability, the function has to go to zero at infinity, which is only possible if

$$\varphi_1(-\infty) = 0 \rightarrow B = 0 \tag{9.9}$$

$$\varphi_3(+\infty) = 0 \rightarrow E = 0 \tag{9.10}$$

Continuity of the function requires

$$\varphi_1(-a) = \varphi_2(-a) \rightarrow Ae^{-\alpha a} = -C\sin\beta a + D\cos\beta a \tag{9.11}$$

$$\varphi_2(a) = \varphi_3(a) \rightarrow Fe^{-\alpha a} = C\sin\beta a + D\cos\beta a \tag{9.12}$$

The function will be twice derivable if the first derivative is also continuous:

$$\varphi_1'(-a) = \varphi_2'(-a) \rightarrow \alpha Ae^{-\alpha a} = \beta C\cos\beta a + \beta D\sin\beta a \tag{9.13}$$

$$\varphi_2'(a) = \varphi_3'(a) \rightarrow -\alpha Fe^{-\alpha a} = \beta C\cos\beta a - \beta D\sin\beta a \tag{9.14}$$

In Eqs (9.11)–(9.14), we have already made use of Eqs (9.9) and (9.10) and abbreviated the first derivative of φ as φ'. From these conditions, it follows[3] that

$$CD = 0 \tag{9.15}$$

So the regularity condition requires either C or D to be zero. Accordingly, there exist two classes of solutions. Depending on whether C or D is set to zero, from Eqs (9.11) and (9.13), on the one hand, and from Eqs (9.14) and (9.12) on the other hand follows that

$$\begin{aligned} C = 0 &\rightarrow \alpha(E) = \beta(E)\tan[a\cdot\beta(E)] \\ D = 0 &\rightarrow \alpha(E) = -\beta(E)\cot[a\cdot\beta(E)] \end{aligned} \tag{9.16}$$

We have already indicated in Section 8.5 that the regularity condition determines the allowed E values and, indeed, we arrived here at equations from which

3 Dividing Eqs (9.13) and (9.14) by Eqs (9.11) and (9.12), respectively, yields

$\alpha = \beta\frac{C\cos\beta a + D\sin\beta a}{-C\sin\beta a + D\cos\beta a}; \quad -\alpha = \beta\frac{C\cos\beta a - D\sin\beta a}{C\sin\beta a + D\cos\beta a}$

and from these follows that

$\frac{C\cos\beta a + D\sin\beta a}{C\sin\beta a - D\cos\beta a} = \frac{C\cos\beta a - D\sin\beta a}{C\sin\beta a + D\cos\beta a}$

Multiplication by the denominators leads to

$$\begin{aligned} &C^2\sin\beta a\cos\beta a + CD\sin^2\beta a + CD\cos^2\beta a + D^2\sin\beta a\cos\beta a \\ &\quad = C^2\sin\beta a\cos\beta a - CD\sin^2\beta a - CD\cos^2\beta a + D^2\sin\beta a\cos\beta a \\ &2CD\underbrace{(\sin^2\beta a + \cos^2\beta a)}_{1} = 0. \end{aligned}$$

E can be obtained. Substituting the definitions of $\alpha(E)$ and $\beta(E)$ from Eqs (9.5) and (9.6),

$$C = 0 \rightarrow (V_0 - E) = E \tan^2\left(a\sqrt{\frac{2m}{\hbar^2}E}\right)$$

$$D = 0 \rightarrow (V_0 - E) = -E \cot^2\left(a\sqrt{\frac{2m}{\hbar^2}E}\right)$$

Unfortunately, these equations cannot be solved analytically. One can, however, get the result numerically or – as we do now – graphically. Multiplication of Eq. (9.16) by a leads to

$$\begin{aligned} C = 0 &\rightarrow a \cdot \alpha(E) = a \cdot \beta(E) \tan[a \cdot \beta(E)] \\ D = 0 &\rightarrow a \cdot \alpha(E) = a \cdot \beta(E) \cot[a \cdot \beta(E)] \end{aligned} \tag{9.17}$$

It should also be considered that, from the definitions of Eq. (9.5), it follows that

$$a^2 \cdot \alpha^2(E) + a^2 \cdot \beta^2(E) = \frac{2m}{\hbar^2} V_0 a^2 \tag{9.18}$$

the right-hand side being a constant. Introducing now the abbreviations

$$X(E) \equiv a \cdot \beta(E); \quad Y \equiv a \cdot \alpha(E); \quad R^2 \equiv \frac{2m}{\hbar^2} V_0 a^2 \tag{9.19}$$

from Eqs (9.17) and (9.18) we obtain two equations for both cases:

$$\begin{aligned} C = 0 &\rightarrow Y = X \tan X; \quad X^2 + Y^2 = R^2 \\ D = 0 &\rightarrow Y = -X \cot X; \quad X^2 + Y^2 = R^2 \end{aligned} \tag{9.20}$$

The two systems of equations are displayed in Figure 9.4. The solutions to both pairs of equations are the points of intersection between the circle ($X^2 + Y^2 = R^2$) and the trigonometric function ($Y = X \cdot \tan(X)$ or $Y = -X \cdot \cot(X)$). One should consider that X and Y must be positive (cf. the definitions in Eqs (9.19) and (9.5)). Now if, for example, the half-width of the well is $a = 2.5$ nm and its height is

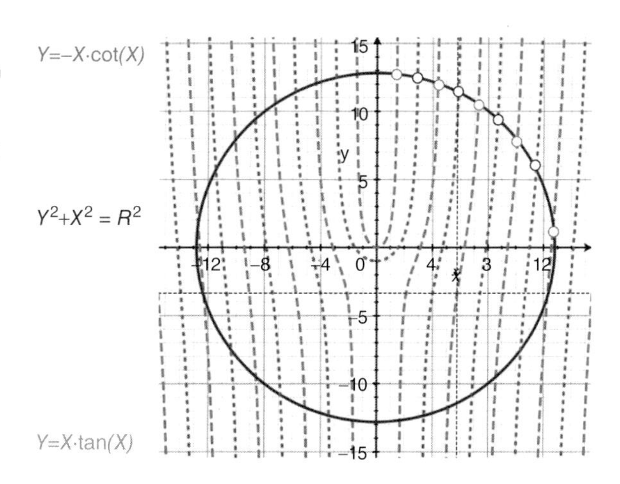

Figure 9.4 Graphic solution for the Schrödinger equation of the quantum well. Color online. Dashed lines correspond to $X \cdot \tan(X)$ and dotted lines to $-X \cdot \cot(X)$ functions.

$V_0 = 1$ eV, then $R^2 \approx 164$, so the intersections of $X \cdot \tan(X)$ with the circle are at $X_{C=0} = 1.5, 4.4, 7.3, 10.1$, and 12.7, while those of $-X \cdot \cot(X)$ are at $X_{D=0} = 2.9$, 5.8, 8.7, and 11.5. As can be seen, these values, up to the last one, increase linearly. According to Eqs (9.5) and (9.19), E is proportional to X^2, so we obtain the following discrete E_n values, at which the solution of the Schrödinger equation is a regular function:

$$
\begin{array}{ll}
C = 0 & D = 0 \\
E_0 = 13.7\,\text{meV} & E_1 = 51.3\,\text{meV} \\
E_2 = 118.0\,\text{meV} & E_3 = 205.1\,\text{meV} \\
E_4 = 324.9\,\text{meV} & E_5 = 461.4\,\text{meV} \\
E_6 = 621.8\,\text{meV} & E_7 = 806.2\,\text{meV} \\
E_8 = 983.2\,\text{meV} &
\end{array}
\tag{9.21}
$$

These are the allowed electron energies in a quantum well of width 5 nm and height 1 eV. Apart from the last two values, the rest are approximately proportional to $(n+1)^2$, where $n = 0, 1, 2, \ldots, 8$. (For $n > 8$ we would be out of the well.) Unfortunately, this derivation does not tell us how the results will change by varying the width or the height: a numerical or graphical solution has to be found in every case separately. However, we can do that quickly by using the Applet http://phet.colorado.edu/en/simulation/bound-states. As can be seen in the snapshots of Figure 9.5, narrower and shallower wells have less allowed energy levels. The differences between the lower levels are not much influenced by the height of the well: these levels are always approximately proportional to the square of the natural numbers. Changing the width, however, influences the proportionality constant considerably (roughly quadratically).

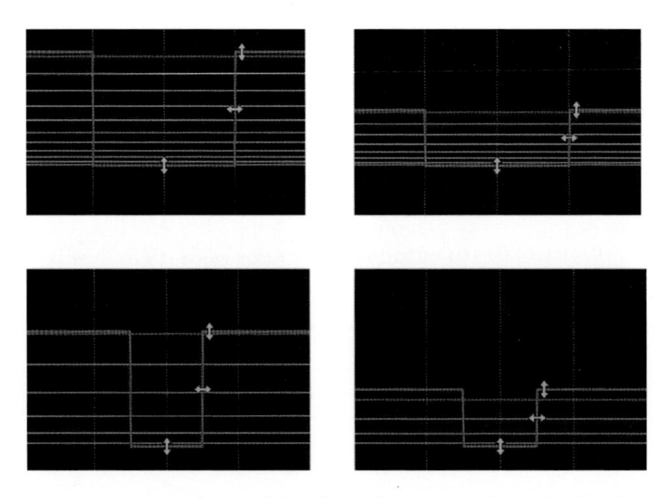

Figure 9.5 Dependence of the allowed energy levels on the height and width of the quantum well. Color online.

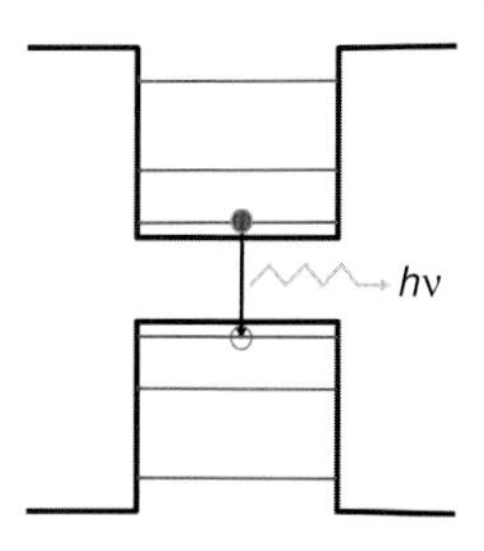

Figure 9.6 Electron and hole levels in the QW-LED.

9.3 Applications in LEDs and in Detectors

An algorithm can easily be written for the procedure described in the previous section and, as with the Applet, one can calculate the allowed energies of electrons and holes for any width $L = 2a$ of the GaAs part and for any Al content x in $Ga_{1-x}Al_xAs$ (which determines V_0). The light-emitting transition occurs between the lowest electron and highest hole state (see Figure 9.6). One can, therefore, design the color by applying appropriate x and L values – at least within the bounds set by the gap between the valence and conduction bands of the middle layer and by the maximal x value at which the same crystal structure between the layers can be retained. GaAs-based LEDs work in the infrared, GaP in the green, and GaN in the blue range of the spectrum. The quantum well (QW) is always built into the n–p junction of a diode.

Besides LEDs and laser diodes, QW diodes are also used in infrared detectors. The electron trapped in the QW can be set free again by absorbing a photon of energy $h\nu > V_0 - E_0$ and can produce a current. By a suitable choice of the QW width, one can sensitize the detector for various frequencies. Microelectronics technology allows a whole array of such highly sensitive detectors to be integrated into one chip. The frequency distribution of the detected radiation can give a clue about its source. (One of the first applications of QWs was the detection of electromagnetic noise, generated by nuclear-powered submarines.)

9.4 Stationary States in a Finite Quantum Well

Knowing the E_n eigenvalues, one can calculate the (discrete) values of the functions $\alpha_n = \alpha(E_n)$ and $\beta_n = \beta(E_n)$ from Eq. (9.5) and, with those, one can obtain the wave function. Using Eqs (9.9) and (9.10), it follows from Eqs (9.6)–(9.8) that

$$\begin{aligned} &\varphi_n(-a > x) = Ae^{\alpha_n x} \\ &\varphi_n(-a < x < a) = C\sin(\beta_n x) + D\cos(\beta_n x) \\ &\varphi_n(x > a) = Fe^{-\alpha_n x} \end{aligned} \tag{9.22}$$

Applying Eqs (9.11) and (9.12) gives

$$C = 0 \rightarrow A_{2l} = F_{2l} = D_{2l}e^{\alpha_{2l}a}\cos\beta_{2l}a \tag{9.23}$$

$$D = 0 \rightarrow A_{2l+1} = -F_{2l+1} = -C_{2l+1}e^{\alpha_{2l+1}a}\sin\beta_{2l+1}a \tag{9.24}$$

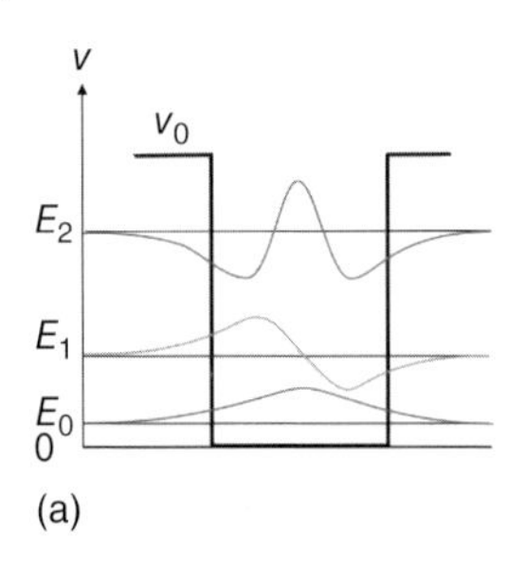

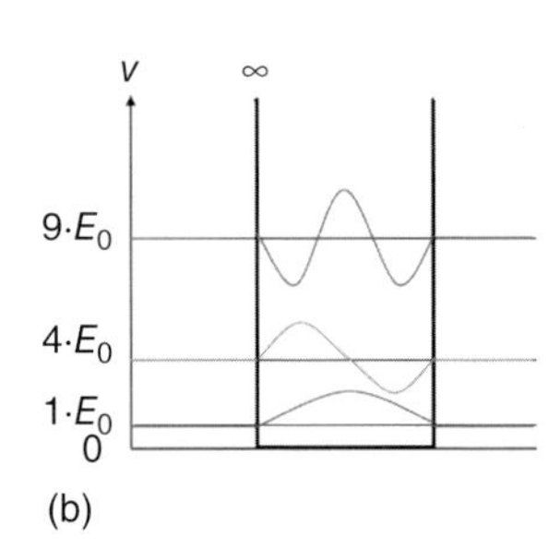

Figure 9.7 Energy eigenstates, with the corresponding energy levels used as x-axis: (a) finite and (b) infinite QW. Color online.

where it was taken into account that in Eq. (9.21) the solutions for $C = 0$ correspond to the even indices $n = 2l$, and those for $D = 0$ to the odd ones $n = 2l + 1$. Equations (9.23) and (9.24) mean that only one parameter, C or D, is left in the composite wave functions of the three regions:

$$\begin{array}{rll} & C = 0 & D = 0 \\ x < -a: & \varphi_{2l}(x) = D_{2l}\cos\beta_{2l}a \cdot e^{\alpha_{2l}(x+a)} & \varphi_{2l+1}(x) = C_{2l+1}\sin\beta_{2l+1}a \cdot e^{\alpha_{2l+1}(x+a)} \\ -a < x < a: & \varphi_{2l}(x) = D_{2l}\cos(\beta_{2l}x) & \varphi_{2l+1}(x) = C_{2l+1}\sin(\beta_{2l+1}x) \\ x > a: & \varphi_{2l}(x) = D_{2l}\cos\beta_{2l}a \cdot e^{-\alpha_{2l}(x-a)} & \varphi_{2l+1}(x) = C_{2l+1}\sin\beta_{2l+1}a \cdot e^{-\alpha_{2l+1}(x-a)} \end{array} \tag{9.25}$$

These remaining parameters can be determined from the condition of square integrability[4]:

$$\begin{aligned} C_{2l+1} &= \left(a + \frac{\sin^2\beta_{2l+1}a}{\alpha_{2l+1}} - \frac{\sin 2\beta_{2l+1}a}{2\beta_{2l+1}}\right)^{-1/2}; \\ D_{2l} &= \left(a + \frac{\cos^2\beta_{2l}a}{\alpha_{2l}} + \frac{\sin 2\beta_{2l}a}{2\beta_{2l}}\right)^{-1/2} \end{aligned} \tag{9.26}$$

The wave function can be depicted using the numerical values of Eq. (9.21). As shown in Figure 9.7a, the results correspond to the expectations of Section 8.5 (cf. Figure 8.4). The functions in Eq. (9.25) are cosines for even and sinus functions for odd n values, with an exponential tail. As expected, n determines the number of nodes. The exponential decay means that there is a finite probability for finding the particle outside the QW!

9.5 The Infinite Quantum Well

As we have seen, for finite V_0, it is not possible to derive an analytical expression for E_n. We have noticed, however, that the numerically obtained E_n values – apart from the ones close to V_0 – seem to follow the formula $E_n = E_0 \cdot (n+1)^2$, more or less independent from the actual value of V_0. So it makes sense to try to derive

4 The appropriate functions should be integrated over the three regions and their sum must give the value of 1.

an approximate formula for the lower levels with $V_0 \to 0$. In that case, α becomes infinite in Eqs (9.6) and (9.8). The exponential tails of the wave function disappear and Eq. (9.7) must yield 0 at $x = \pm a$:

$$C\sin(+\beta a) + D\cos(+\beta a) = C\sin\beta a + D\cos\beta a = 0$$
$$C\sin(-\beta a) + D\cos(-\beta a) = C\sin\beta a - D\cos\beta a = 0 \tag{9.27}$$

If neither C nor D is zero, these two equations cannot be satisfied at the same time. Therefore, we have again two classes of solutions:

$$C = 0 \;\to\; \varphi = D\cos(\beta x) \tag{9.28}$$

$$D = 0 \;\to\; \varphi = C\sin(\beta x) \tag{9.29}$$

From the condition that both must go to zero for $x = \pm a$, we obtain

$$C = 0 \;\to\; \cos(\beta a) = 0 \;\to\; \beta_{2l+1}a = (2l+1)\frac{\pi}{2} \;\to\; \varphi_{2l+1} = D\cos\left[(2l+1)\frac{\pi}{2a}x\right] \tag{9.30}$$

$$D = 0 \;\to\; \sin(\beta a) = 0 \;\to\; \beta_{2l}a = (2l)\frac{\pi}{2} \;\to\; \varphi_{2l} = C\sin\left[(2l)\frac{\pi}{2a}x\right] \tag{9.31}$$

and the normalization gives $C = D = \sqrt{2/L}$. Since Eq. (9.31) gives the trivial solution $\varphi_{2l}(x) = 0$ for $l = 0$, the ground state is supplied by Eq. (9.30) for $l = 0$, with no nodes. The solutions for $L = 2a$ can then be summarized as

$$\varphi_{n=2l}(x) = \sqrt{\frac{2}{L}}\cos\left[(n+1)\pi\frac{x}{L}\right]; \quad \varphi_{n=2l+1}(x) = \sqrt{\frac{2}{L}}\sin\left[(n+1)\pi\frac{x}{L}\right] \tag{9.32}$$

From Eqs (9.30) and (9.31) with the definition of β in Eq. (9.5) follows the analytical expression of the energy eigenvalues:

$$E_n = \frac{h^2}{8mL^2}(n+1)^2,\; n = 0,1,\ldots \tag{9.33}$$

As can be seen, for the infinite quantum well, we have indeed arrived at a formula containing $(n+1)^2$, and E_0 is determined by the square of the width. As our numerical solutions in Section 9.2 demonstrate, *this formula is a good approximation for the lower eigenvalues in finite wells* as well.

Note 1: As shown by Eq. (9.33), energy of the ground state (i.e., the minimal energy) is not zero, although at the bottom of the well $V = 0$. This means that even at a (hypothetical) temperature of 0 K, the kinetic energy cannot be zero. As shown in Figure 9.7b, the ground state is a standing cosine wave. Referring to 0 K, this is called the *zero-point vibration*. It is a consequence of Heisenberg's uncertainty principle: if the kinetic energy was zero, the electron would stand still and both Δx and Δp_x would simultaneously be zero.

Note 2: According to Eq. (9.33), the allowed energies in the QW are quantized. The energy difference between states with different n depends on the mass of the particle m. At, for instance, $m = 1$ kg, the energy differences are about 10^{30} times smaller than for the electron. In other words, for such a mass, the levels would follow each other quasi-continuously. In our size range, quantization cannot be noticed.

9.6 Comparison to a Classical Particle in a Box

The three-dimensional quantum well (quantum box) localizes the electron in a small area. Still, the energy eigenstate cannot be at the same time an eigenstate of position as well, since the kinetic energy operator ($\sim d^2/dx^2$) is not commutative with the position operator ($x\cdot$). The probability density is given by the absolute square of Eq. (9.32), and it is a symmetric function of x for both even and odd n. Consequently, the expectation value of the measured position in every state is[5]

$$\langle x\rangle = \int_{-L/2}^{L/2} x\varphi_n^2(x)dx = 0 \tag{9.34}$$

that is, the middle of the box, also for odd n, even though, for example, $|\varphi_{n=2l+1}(0)|^2 = 0$! Using Eq. (9.34) in the definition of the uncertainty, $\Delta x = \sqrt{\langle x^2\rangle - \langle x\rangle^2}$, we obtain

$$\Delta x = L\sqrt{\frac{1}{12} \mp \frac{1}{2(n+1)^2\pi^2}} \tag{9.35}$$

where the minus sign holds for even and the plus for odd n values.

The energy eigenstates are no momentum eigenstates either. The expectation value of the momentum is[6]

$$\begin{aligned}\langle p_x\rangle &= \int_{-L/2}^{L/2} \varphi_n(x)\frac{\hbar}{i}\frac{d}{dx}\varphi_n(x)dx \\ &= \frac{2\hbar(n+1)\pi}{iL^2}\int_{-L/2}^{L/2} \sin\left[\frac{(n+1)\pi x}{L}\right]\cos\left[\frac{(n+1)\pi x}{L}\right]dx = 0\end{aligned} \tag{9.36}$$

as expected for a standing wave. The uncertainty of the momentum is

$$\Delta p_x = \sqrt{\langle p_x^2\rangle} = \sqrt{\int_{-L/2}^{L/2} \varphi_n(x)\left(\frac{\hbar}{i}\right)^2\frac{d^2}{dx^2}\varphi_n(x)dx} = (n+1)\frac{\hbar\pi}{L} \tag{9.37}$$

With $\langle x\rangle = 0$ and $\langle p_x\rangle = 0$, the particle in the box is closest to the classical point mass; however, as shown by Eqs (9.35) and (9.37), the uncertainties are substantial. For example, in the ground state, $\Delta x = \pm 0.36a$. From these equations it follows, too, that even the smallest value of the product of uncertainties, $\Delta p_x \Delta x = \hbar\sqrt{(\pi^2/12) - (1/2)} = 0.568\hbar$, is higher than $\hbar/2$, in agreement with Heisenberg's uncertainty principle.

5 Since φ_n^2 is always an even function and x is uneven, the integral of the product must be zero.
6 Because, here again, an even function is integrated symmetrically to the origin of the coordinate system.

Summary in Short

- The allowed energies of a particle in an *infinite* quantum well are quantized according to $E_n = (h^2/(8mL))(n+1)^2$ and can be tuned by the width L of the well. This expression can also be used as an approximation for the lower states in a *finite* quantum well. In agreement with the uncertainty principle, the energy of the ground state ($n = 0$) is not zero.
- The stationary states in the infinite quantum well are standing waves with the position-dependent parts: $\sqrt{2/L}\cos[(2l)\pi(x/L)]$ and $\sqrt{2/L}\sin[(2l+1)\pi(x/L)]$, going to zero at the wall of the well. In contrast to that, the wave function decays exponentially beyond the boundaries of the finite quantum well. As a consequence, the probability of finding the electron outside the well is not zero.
- The condition of regularity (square integrability) leads to quantized *energy levels* for the particle in a finite quantum well too. The number and energy distance of the levels can be tuned by the height and width of the well. The higher and wider the well, the more levels are allowed. Decreasing the height or increasing the width leads to smaller differences between the levels. For very wide potential wells or for large particle masses, the distribution of levels becomes quasi-continuous.

9.7 Questions and Exercises

Problem 9.1 Two energy eigenstates in an infinite quantum well have two and four nodes, respectively. If the width of the well is $L = 10$ nm, what is the difference in the emitted photon energy if the electron returns to the ground state from one or the other state?

Problem 9.2 The ground state of an electron in an LED can be given as $E_n^e = E_C - (h^2/(8mL^2))(n+1)^2$, where E_C is the energy of the conduction band edge. Assuming $E_n^h = E_V + (h^2/(8mL^2))(n+1)^2$ for the holes (with E_V as the valence band edge) and a gap of $E_C - E_V = 1.4$ eV, how should the width L be chosen if we want emission in the red with $\lambda = 700$ nm? Can such an LED be realized technically?

Problem 9.3 An element of a detector array has to react for the wavelength 3000 nm. Assuming a quantum well width of 10 Å and that Eq. (9.33) can be applied for the ground state, how deep should the well be?

Problem 9.4 Calculate the expectation value of the energy in the ground state of a quantum well, making use of the fact that the bottom of the well is at $V = 0$, so the energy is purely kinetic and determined by $\langle p^2 \rangle$. Compare the result with the energy of the zero-point vibration in Eq. (9.33)! (Hint: $\int_{-L/2}^{+L/2} \cos^2\left(\frac{\pi}{L}x\right) dx = \frac{L}{2}$.)

Problem 9.5 Try to derive Eq. (9.33) from the interference condition for a string of length L! Use Eq. (8.10) for the momentum.

10

The Tunnel Effect and Its Role in Electronics

In this chapter...
We investigate the consequences of the fact (found in Chapter 9) that the wave function is decaying exponentially into the wall of a finite quantum well. According to Eq. (9.25), the higher the top of the well V_0 is above the total energy E, and the bigger the mass m is, the bigger is the decay constant, $\alpha = \sqrt{2m(V_0 - E)/\hbar^2}$, and the faster is the decay. Reversely, this means that a particle of small mass can penetrate deep into the wall of a not too deep well. If the wall is thin, the absolute square of the wave function can still be substantial on the other side of the wall. In other words, the probability density of finding the particle outside the well will be non-negligible.[1] This phenomenon can, in fact, be observed. An electron can go through a potential wall even if it cannot "jump" over it (i.e., if $E < V_0$). This is called the *tunnel effect* – even though no tunnel exists in the wall. It is a consequence of the wave nature of the electrons and of the uncertainty principle.

10.1 The Scanning Tunneling Microscope

The tunnel effect is spectacularly demonstrated by the scanning tunneling microscope (STM), which allows the imaging of a metal surface with atomic resolution. In the STM, a very sharp needle is driven over the surface piezoelectrically, without touching it. Although the air or vacuum gap between the needle and the surface is insulating (corresponds to a potential wall), an electron current can flow from the surface to the needle if applying a voltage U, even with qU being smaller than the work function (i.e., the energy needed for an electron to leave the metal). This tunneling current is proportional to e^{-d}, where d is the distance between the needle and the surface, that is, the width of the potential wall. The exponential dependence makes the current very sensitive to the height differences in the surface, so the current can be used to map the surface with a height resolution less than 1 nm, as shown in Figure 10.1. In the following sections, we

1 This probability, however, decreases with increasing mass and becomes negligible for macroscopic bodies.

Essential Quantum Mechanics for Electrical Engineers, First Edition. Peter Deák.

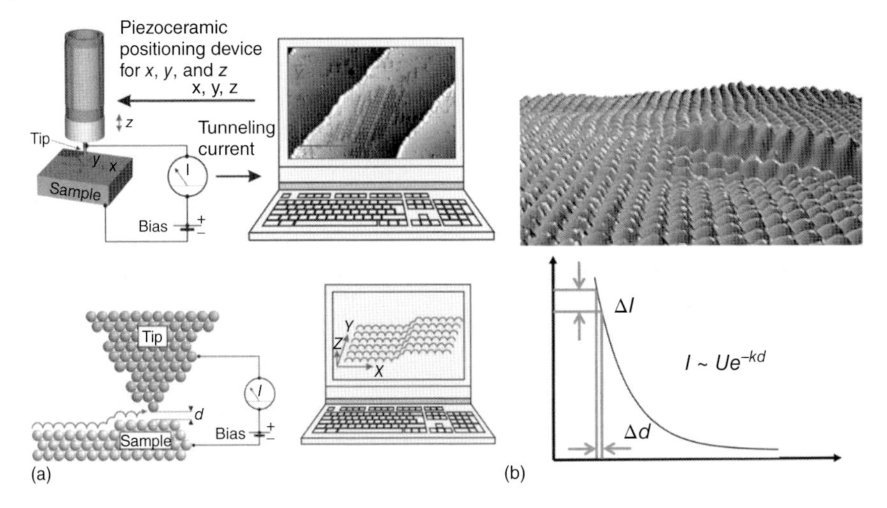

Figure 10.1 Schematic representation of a scanning tunneling microscope. Color online. (a: Reproduced with permission of Rene Pascal, www.beugungsbild.de. b: Reproduced with permission of Mattias Bode, www.physik.uni-wuerzburg.de University of Wuerzburg.)

consider this distance dependence of the tunneling current in detail and consider the significance of the tunnel effect in various areas, such as electron emission from a cathode, leakage currents in integrated circuits (ICs), electric breakdown, quantum-cascade lasers, and flash memories, etc.

10.2 Electron at a Potential Barrier

Let us consider the one-dimensional case of a potential wall shown in Figure 10.2, which is approached from the left by an electron with a given energy $E < V_0$.

Here too, the time-independent Schrödinger equation can be constructed separately in three regions:

$$x < 0 : \; -\frac{\hbar^2}{2m}\frac{d^2\varphi_1}{dx^2} = E\varphi_1 \;\rightarrow\; \frac{d^2\varphi_1}{dx^2} = -k^2\varphi_1 \tag{10.1}$$

$$0 < x < d : \; -\frac{\hbar^2}{2m}\frac{d^2\varphi_2}{dx^2} + V_0\varphi_2 = E\varphi_2 \rightarrow \frac{d^2\varphi_2}{dx^2} = \frac{1}{4b^2}\varphi_2 \tag{10.2}$$

$$x > d : \; -\frac{\hbar^2}{2m}\frac{d^2\varphi_3}{dx^2} = E\varphi_3 \;\rightarrow\; \frac{d^2\varphi_3}{dx^2} = -k^2\varphi_3 \tag{10.3}$$

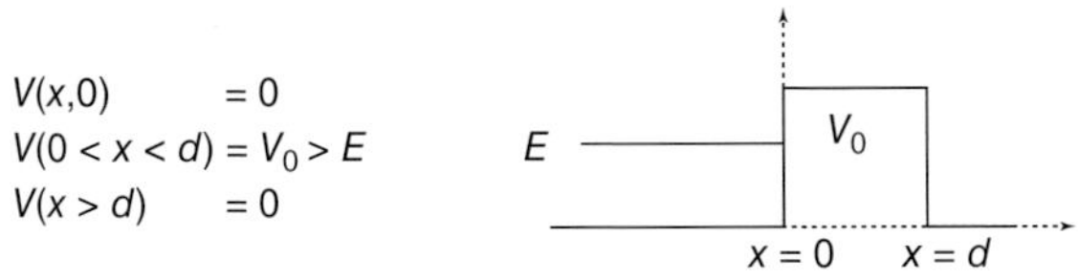

$V(x,0) = 0$
$V(0 < x < d) = V_0 > E$
$V(x > d) = 0$

E
V_0
x = 0   x = d

Figure 10.2 One-dimensional potential wall.

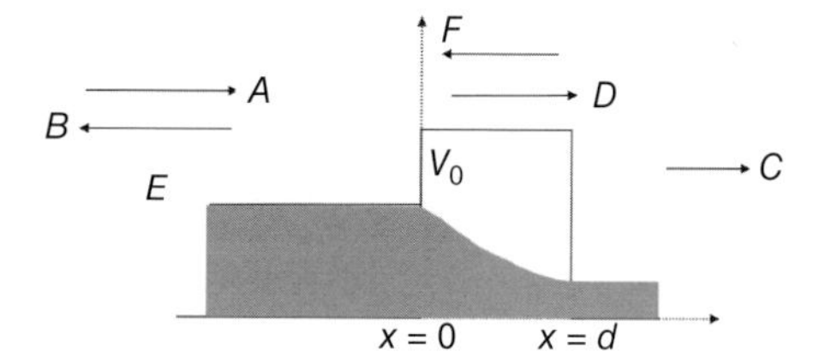

Figure 10.3 Electron transmission at a potential wall (cf. Eq. (10.6)).

where we have introduced the following abbreviations:

$$k^2 \equiv \frac{2mE}{\hbar^2}; \; \frac{1}{4b^2} \equiv \frac{2m(V_0 - E)}{\hbar^2} \tag{10.4}$$

$$\gamma \equiv 2bk = \sqrt{\frac{E}{V_0 - E}} \tag{10.5}$$

As can be seen by substituting into Eqs (10.1)–(10.3), the solution in the three regions can be written as

$$\varphi_1 = Ae^{ikx} + Be^{-ikx}; \; \varphi_2 = De^{x/2b} + Fe^{-x/2b}; \; \varphi_3 = Ce^{ikx} \tag{10.6}$$

The two terms of φ_1 in the region $x < 0$ represent (after multiplication with the time factor $e^{-i(E/\hbar)t}$) a harmonic wave coming from the left with amplitude A, and another with amplitude B propagating in the opposite direction. These correspond to the incoming and the reflected electron, respectively (see Figure 10.3). The function φ_3 in the region $x > d$ represents a harmonic wave propagating to the right with amplitude C, that is, the electron that has tunneled through the wall. (No electron is coming from the right, so no e^{-ikx} is necessary here.) The function φ_2 describes the electron in the wall.

The tunneling current is proportional to the transmission rate (or tunneling probability):

$$T = \left|\frac{C}{A}\right|^2 \tag{10.7}$$

In order to calculate T, we have to determine the coefficients in Eq. (10.6) from the regularity condition of Postulate 1, by requiring at the boundaries:

$$\text{Continuity:} \;\; \varphi_1(0) = \varphi_2(0); \; \varphi_2(d) = \varphi_3(d) \tag{10.8}$$

$$\text{Continuous derivative:} \;\; \varphi_1'(0) = \varphi_2'(0); \; \varphi_2'(d) = \varphi_3'(d) \tag{10.9}$$

where, again, the abbreviation $\varphi' \equiv d\varphi/dx$ was used. These equations lead to[2]

---

2 From Eqs (10.8) and (10.9) follows

$$A + B = D + F; \; De^{d/2b} + Fe^{-d/2b} = Ce^{ikd}$$
$$A - B = \frac{1}{i\gamma}(D - F); \; De^{d/2b} - Fe^{-d/2b} = i\gamma Ce^{ikd}$$

Adding up the two equations on the left and using Eq. (10.5) gives

$$2A = \left(1 + \frac{1}{i\gamma}\right)D + \left(1 - \frac{1}{i\gamma}\right)F$$

The other two equations give, after addition and after subtraction,

$$D = \frac{1}{2}(1 + i\gamma)e^{-d/2b}Ce^{ikd}; \; F = \frac{1}{2}(1 - i\gamma)e^{d/2b}Ce^{ikd}$$

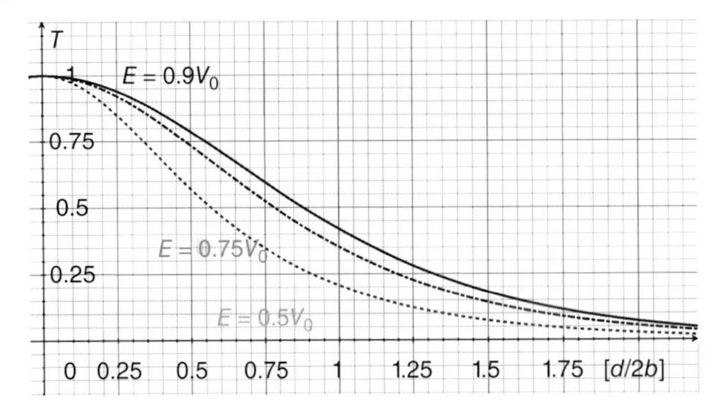

Figure 10.4 Dependence of the electron transmission T on the width of the potential wall d (in $1/2b$ units) for different E/V_0 ratios. Color online.

$$T = \left|\frac{C}{A}\right|^2 = \left[1 + \frac{V_0^2}{4E(V_0 - E)}\sinh^2\left(\frac{d}{2b}\right)\right]^{-1} \tag{10.10}$$

This general expression for the tunneling probability, which for given E is a function of the V_0 height and d width of the potential wall, is displayed in Figure 10.4 for various E/V_0 ratios. As can be seen, the curves show an exponential decay for $d/2b > 1$.

Substituting these results for D and F into the equation for A above leads to

$\frac{A}{C}e^{-ikd} = \frac{1}{4}\left[\left(1 + \frac{1}{i\gamma}\right)(1 + i\gamma)e^{-d/2b} + \left(1 - \frac{1}{i\gamma}\right)(1 - i\gamma)e^{d/2b}\right]$

The expression in the parentheses can be written as

$\left(1 \pm \frac{1}{i\gamma}\right)(1 \pm i\gamma) = 1 \pm i\gamma \pm \frac{1}{i\gamma} + 1 = 2 \pm i\frac{\gamma^2-1}{\gamma}$

and with that, using the definitions

$\cosh(x) \equiv \frac{e^x+e^{-x}}{2};\ \sinh(x) \equiv \frac{e^x-e^{-x}}{2}$

we obtain

$\frac{A}{C}e^{-ikd} = \frac{1}{4}\left[\left(2 + i\frac{\gamma^2-1}{\gamma}\right)e^{-d/2b} + \left(2 - i\frac{\gamma^2-1}{\gamma}\right)e^{d/2b}\right] = \frac{e^{d/2b}+e^{-d/2b}}{2} - i\left(\frac{\gamma^2-1}{2\gamma}\right)\frac{e^{d/2b}-e^{-d/2b}}{2} =$
$\cosh\left(\frac{d}{2b}\right) - i\left(\frac{\gamma^2-1}{2\gamma}\right)\sinh\left(\frac{d}{2b}\right)$

The absolute square of this equation is

$\left|\frac{A}{C}e^{-ikd}\right|^2 = \left|\frac{A}{C}\right|^2 = \cosh^2\left(\frac{d}{2b}\right) + \left(\frac{\gamma^2-1}{2\gamma}\right)\sinh^2\left(\frac{d}{2b}\right)$

because $e^{-ikd}e^{ikd} = 1$ and the absolute square of a complex number, $z = x + iy$, is $x^2 + y^2$.

From here the transmission coefficient is

$\left|\frac{C}{A}\right|^2 = \left[\cosh^2\left(\frac{d}{2b}\right) + \left(\frac{\gamma^2-1}{2\gamma}\right)\sinh^2\left(\frac{d}{2b}\right)\right]^{-1}$

Since $\cosh^2(x) = 1 - \sinh^2(x)$, and from Eq. (10.5)

$\left(\frac{\gamma^2-1}{2\gamma}\right)^2 = \frac{V_0^2}{4E(V_0-E)}$

Equation (10.10) follows.

Equation (10.4) defines the *penetration depth b* as

$$b = \frac{\hbar}{\sqrt{8m(V_0 - E)}} \tag{10.11}$$

In the case of an electron, b in (Å) can be given approximately as a function of $\Delta E \equiv E - V_0$ in (eV), as

$$b(\text{Å}) \approx \frac{1}{\sqrt{\Delta E(\text{eV})}} \tag{10.12}$$

This means that for a typical value of $\Delta E \approx 1$ eV, $b \approx 1$ Å. Since the atomic distances in a crystal are usually 2.5 Å or more, $d/2b$ is >1 in every practical case. So we can apply an exponential approximation for Eq. (10.10). Utilizing the fact that for large x values $\sinh(x) \approx e^{ix/2}$, we get

$$T \approx 16\frac{E(V_0 - E)}{V_0^2} \exp\left(-\frac{d}{b}\right) \tag{10.13}$$

This is the exponential dependence of the tunneling rate on the width of the potential wall, as seen in STM (Fig.10.1). This approximation is practically always applicable for electrons. Here, we can also see the meaning of the penetration depth: for $d = b$ the incoming current density, $j_0 \sim T$, falls to j_0/e beyond the wall.

Our derivation so far has assumed the incoming electron to be in an energy eigenstate. More realistic but also mathematically more demanding is to start with a wave packet. The tunneling of a wave packet (as well as that of a harmonic wave) can be studied, however, with the help of the applet http://phet.colorado.edu/en/simulation/quantum-tunneling. Figure 10.5 shows snapshots as time progresses.

As can be seen, after the packet collides with the wall, it is partly reflected but part of it propagates further on the other side, away from the wall. Of course, this does not mean that the electron, which it describes, is being divided, only that the probability of finding the electron after the collision on either side of the wall will be (under the circumstances) the same. An attempt to detect the electron would localize it at once on one or the other side.

By setting $E > V_0$ in the applet, one can see that even in this case T could be less than 1. Therefore, in contrast to the prediction of classical physics, an electron can be reflected by a potential wall even if its energy would be sufficient to go over it!

10.3 Field Emission, Leakage Currents, Electrical Breakdown, Flash Memories

Procedures, based on ideas similar to that of the STM, play a great role in the nanotechnology presently under development. However, the tunnel effect has great significance already from the viewpoint of today's electrical engineering. In cathode emission (as in the cathode ray tube of a traditional television apparatus or in the discharge tube of fluorescent lamps), a potential threshold has to be overcome by the electrons to be able to leave the cathode. Let us neglect now the effect of

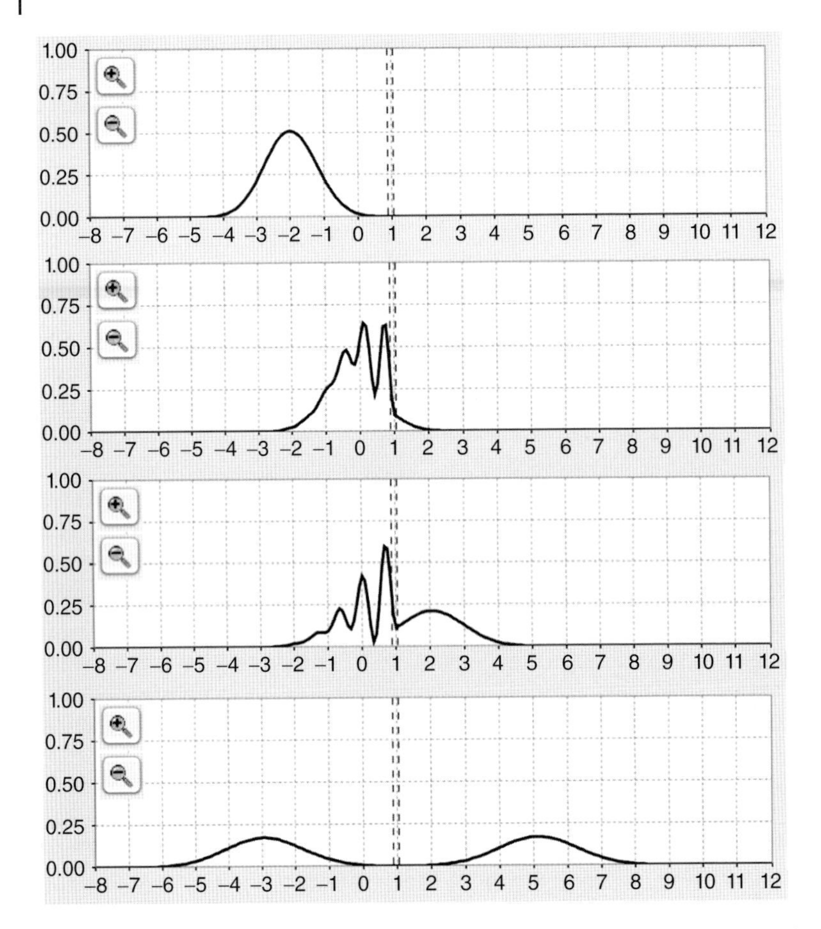

Figure 10.5 Wave packet at the potential wall (dashed lines in the middle). (Snapshots from the applet http://phet.colorado.edu/en/simulation/quantum-tunneling.)

heating of the cathode, and consider the effect of the bias U, applied between the electrodes.[3] In other words, we assume pure *field emission*.

As shown in Figure 10.6, constant bias corresponds to a homogeneous **E**-field of strength $|\mathbf{E}| = U/l$, where l is the distance between the electrodes. As it is known from electromagnetics, the **E**-field cannot penetrate into metal electrodes but outside them it adds

$$V(x) = -q|\mathbf{E}|x \tag{10.14}$$

to the potential energy of the electrons (dashed line in Figure 10.6). Let us set the origin of the coordinate system with $x = 0$ at the surface of the cathode and the energy $V = 0$ at the bottom of the potential well, representing the cathode.

If the work function of the cathode is W (cf. Section 3.1), the electron with the highest energy, e_m, has to cross a triangular potential wall, as shown in Figure 10.6, to leave the electrode. Following from Eq. (10.14), the basis of the triangle has

3 Actually, a bias is also applied in the STM. We have only neglected it in Section 10.3.

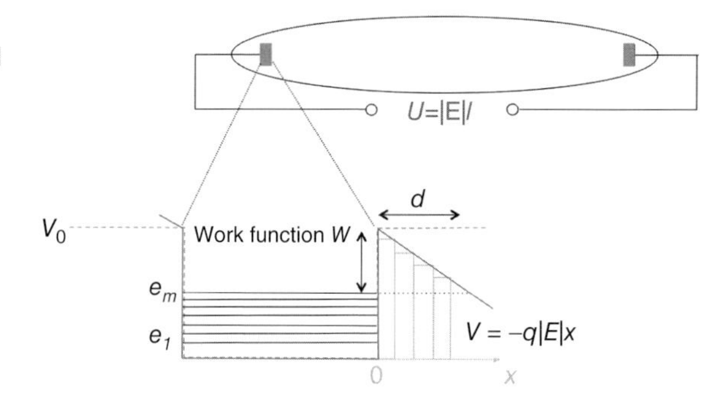

Figure 10.6 Potential in the vicinity of the cathode in field emission.

the width of $d = W/(q|\mathbf{E}|)$. One can approximate this triangle with a series of thin rectangular potential walls of width Δx_i $(i = 1, 2, 3, \ldots)$. For $\Delta x_i \to 0$, the approximation is exact. The probability of tunneling through the triangular wall is the product of the probabilities for tunneling through the individual rectangular walls. For the latter, we can apply Eq. (10.13) and so[4]

$$T = \prod_i 16\frac{e_m[V(x_i) - e_m]}{V^2(x_i)} \exp\left(-\frac{2\sqrt{2m[V(x_i) - e_m]}}{\hbar}\Delta x_i\right) \tag{10.15}$$

Since the prefactor is roughly proportional to $1/x$, it varies much more slowly than the exponential function behind it. Therefore, to a first approximation, we can substitute the prefactor with a constant T_0. Considering that the current density j is proportional with the transmission rate T, we obtain the *Fowler–Nordheim tunnel formula*

$$j = j_0 \exp\left(-\frac{\sqrt{32m}}{3\hbar}\frac{W^{3/2}}{q|\mathbf{E}|}\right) \tag{10.16}$$

where j_0 is the incoming current density.[5]

Actually, this formula should be used for the STM, but it is applicable also in the simple case of an electric plug, which is a metal (usually copper or aluminum) with a thin, insulating oxide layer on its surface, which represents a potential wall. The tunneling current has a special significance in metal–oxide–semiconductor (MOS) field effect transistors (FETs), which are the basic elements of ICs. As shown in Figure 10.7, a MOS-FET consists of a metal electrode (M), called "gate,"

4 As a reminder, $\prod_i X_i = X_1 \cdot X_2 \cdot X_3 \cdot \cdots$.

5 Since with the help of Eq. (10.14)

$V(x_i) - e_m = (e_m + W - q|\mathbf{E}|x_i) - e_m = q|\mathbf{E}|d\left(1 - \frac{x_i}{d}\right) = W\left(1 - \frac{x_i}{d}\right)$

we obtain

$$\lim_{\Delta x_i \to 0} T \approx T_0 \lim_{\Delta x_i \to 0} \prod_i \exp\left(-\sqrt{8mW\left(1 - \frac{x_i}{d}\right)}\frac{\Delta x_i}{\hbar}\right) = T_0 \exp\left(-\frac{\sqrt{8mW}}{\hbar}\int_0^d \left(1 - \frac{x}{d}\right)^{1/2} dx\right)$$

where the identity $\prod_i e^{X_i} = e^{\sum_i X_i}$ was used, as well as the fact that the summation can be replaced by an integral in the limit if $\Delta x_i \to 0$. The integral has the value of $2d/3$ and, with $d = W/(q|\mathbf{E}|)$, Eq. (10.16) follows.

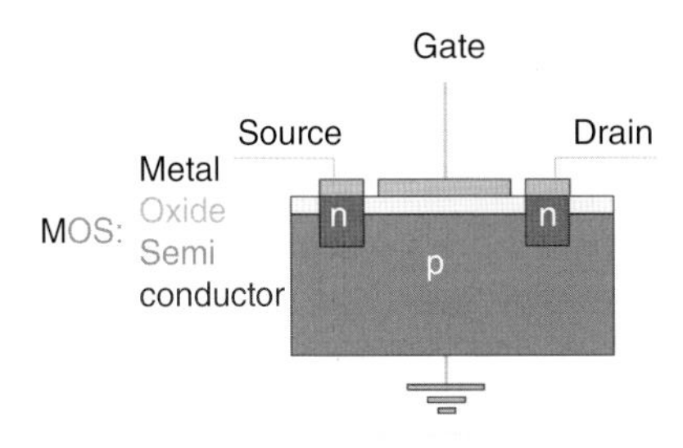

Figure 10.7 Scheme of a MOS-FET. Color online. (The semiconductor has different electrical properties in the regions marked n and p.)

which is separated by an oxide layer (O) from a grounded semiconductor (S). A current in the semiconductor layer between "source" and "drain" can only flow if applying a critical bias, U_G, between the metal electrode and the ground. Of course, current must not flow between the gate and the semiconductor, and that is why the insulating oxide layer is needed.

The simple "yes" or "no" current state of a MOS-FET allows to build logic gates, and ICs contain millions of them. Miniaturization (increasing scale of integration) has made the MOS-FETs ever smaller and the oxide layer ever thinner. Therefore, the leakage current due to tunneling between the gate and the semiconductor is increasing, and that will stop further shrinkage of the MOS-FET size in the foreseeable future. Presently, all ICs consist of such MOS-FET elements, but in the future new solutions must be found to further increase the speed and capacity of electronic devices.

Actually, the tunnel effect can also be used for switching as, for example, in the *Zener diodes*, where electric breakdown under reverse bias is utilized. Tunneling is also often responsible for the electric breakdown of insulators in general. In Section 3.1, at the photovoltaic effect, it has already been mentioned that the allowed energies of electrons not only in semiconductors but also in insulators form two quasi-continuous bands along the energy scale. The lower, so-called valence band is completely filled (every state is occupied) and cannot contribute to the total current. The upper one, called conduction band, is empty, so current can only flow if electrons are promoted from the valence band to the conduction band (by photoexcitation or thermally). The minimum energy needed corresponds to the energy gap between the two bands, E_g. In most insulators and semiconductors (also in silicon), E_g is much bigger than the thermal energy at the melting point and that is the reason for the insulating behavior.[6] However, if a homogeneous field of strength $\mathbf{E}$ is switched on along, for example, the x-direction, the potential energy $V(x) = -q|\mathbf{E}|x$ has to be added to the energy of the electrons. As shown in Figure 10.8, if the field strength achieves a critical value, electrons from the valence band can tunnel through the depicted triangular potential wall into the conduction band, and the material will be conductive. Replacing the work function W in the Fowler–Nordheim tunnel formula, Eq. (10.16), by the band gap E_g, the leakage current can be calculated. The breakdown field strength corresponds to a critical leakage current.[7]

6 One has to dope silicon with other atoms to achieve reasonable conductivity at room temperature.
7 N.B.: The breakdown has another reason too. With increasing kinetic energy, obtained from the field, the electrons are scattered more strongly and inelastically by the atoms, until the crystal

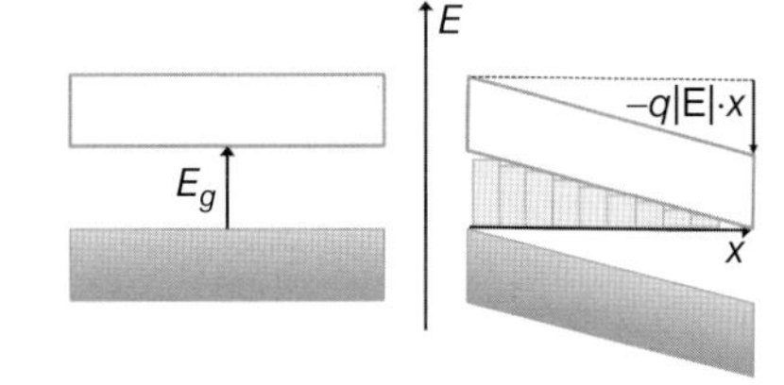

Figure 10.8 Electric breakdown of an insulator due to the tunnel effect. The triangular potential wall, arising after the voltage is applied, is depicted similarly to Figure 10.6. Color online.

The tunnel effect has another very important application in solid-state data storage devices (or flash memories). This is investigated in Problem 10.3.

10.4 Resonant Tunneling, Quantum Field Effect Transistor, Quantum-Cascade Lasers

Considering the limiting effect of the tunnel currents on the further miniaturization of present MOS-FET devices, one should think of alternative ideas that actually make use of tunneling. One possibility is offered by the *resonant tunnel effect*. It can be shown[8] that the tunneling rate through two thin potential walls, which form a quantum well, increases enormously if the incoming electron has an energy coinciding with one of the allowed energy levels in the well (cf. Eq. (9.21) or Eq. (9.33)). With the applet https://phet.colorado.edu/en/simulation/quantum-tunneling, one can follow this in animation. Figure 10.9 shows snapshots from the applet for harmonic waves of various energies coming from the left. As can be seen, the tunneling is substantial (the amplitude right of the well is comparable with that on the left) for certain energies only. At these energies, standing waves form within the well, as expected from Eq. (9.25) (cf. Figure 9.7), and such energy values follow each other roughly as $\sim (n+1)^2$. This means that the electron can tunnel through both walls if its energy corresponds to one of the allowed energy eigenvalues within the well. This is the resonant tunnel effect.

The resonant tunneling can be utilized, for example, in a *quantum FET* (http://sandia.gov/media/quantran.htm). One can design a double potential wall for electrons by sandwiching, for example, GaAs and $Ga_{1-x}Al_xAs$ layers as shown in Figure 10.10. Applying a gate voltage to the middle layer, the allowed energy levels in the quantum well can be shifted ($V = -qU_G$ is added to E_n). For a critical value of U_G, one of the E_n eigenvalues becomes equal to the energy of the incoming electrons and resonant tunneling allows a current to flow. The device is then in the "ON" mode. As can be seen in Figure 10.9, the tunneling rate falls off rapidly even for a small change of U_G relative to the critical bias, and the device is switched in the "OFF" mode. Of course, the gate still has to be isolated from the semiconductor, but due to the sensitivity of the resonant tunneling, the necessary gate voltage is much lower than in the traditional MOS-FETs, so a very thin insulator is sufficient. Such devices are still in the development phase (see http://www.digitaltrends.com/computing/analyst-predicts-intel-newmethods-moores-law/

structure deteriorates. However, this happens mostly for stronger fields than the breakdown through tunneling currents.

8 See the derivation in chapter 10.4.1.

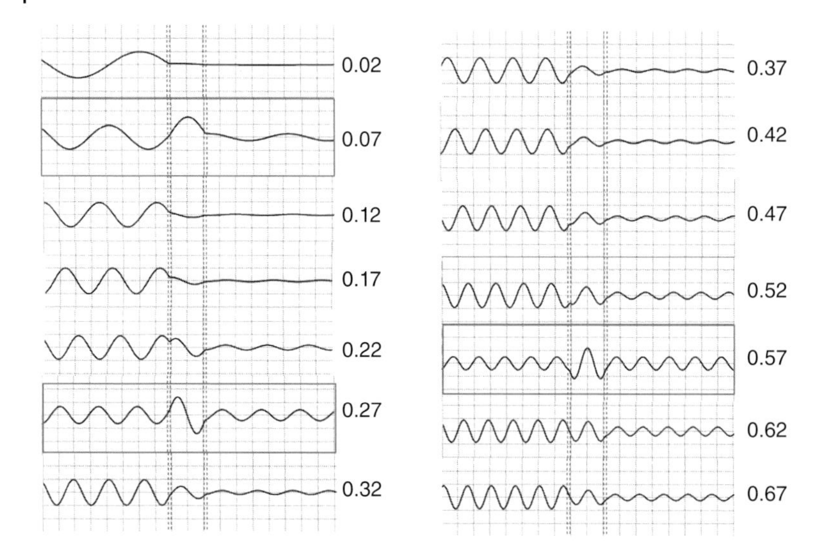

Figure 10.9 Penetration of harmonic waves, with energy increasing in 0.05 eV steps, into a quantum well (width 2 nm, height 1.5 eV, wall thickness 0.2 nm). Dashed lines show the position of the potential walls. (Snapshots from the applet https://phet.colorado.edu/en/simulation/quantum-tunneling).

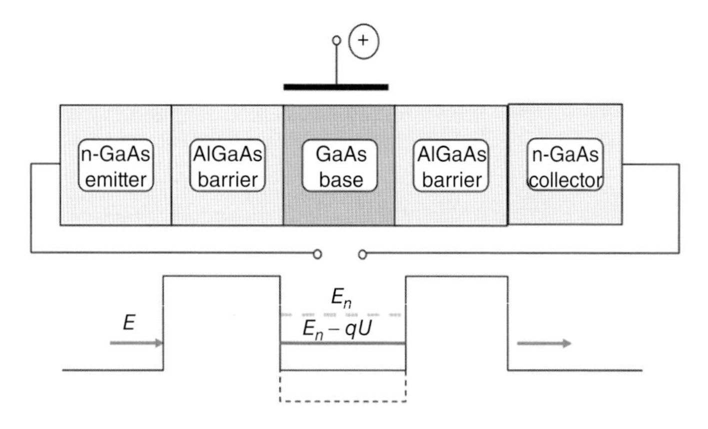

Figure 10.10 Schematic view of a quantum field effect transistor. Color online.

and http://www.theregister.co.uk/2005/12/07/intel_qinetiq_quantum_well/). Actually one of the main problems is the price of depositing an appropriate insulator layer. (In silicon-based MOS-FETs, the insulating oxide layer grows on silicon by itself upon exposure to oxygen.)

Another application of the resonant tunneling, the *quantum-cascade laser* is already in the market and allows for the production of high light power at variable wavelength with a semiconductor diode. The cascade laser contains

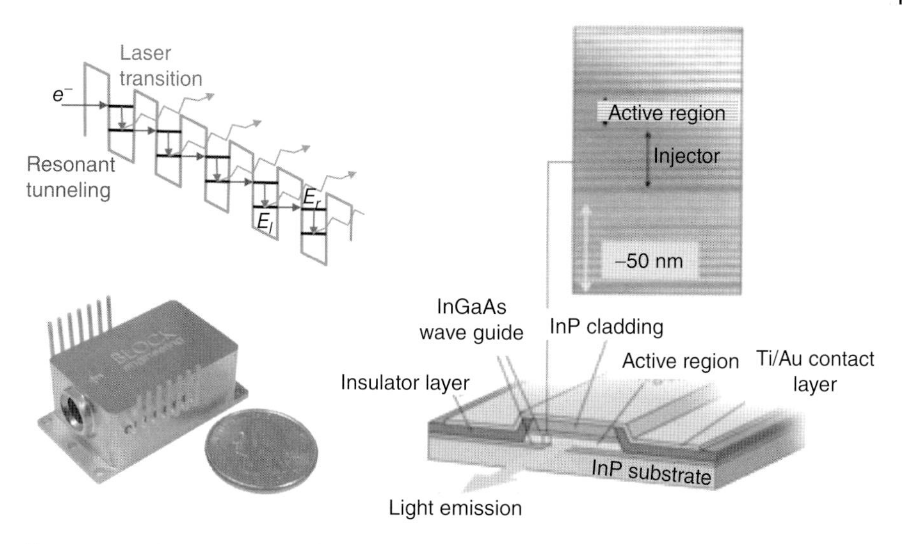

Figure 10.11 Operation principle and structure of a cascade laser (reproduces by the kind permission of Conrad Holton) and a commercial tunable cascade laser (Block engineering's Mini-QCL® https://www.laserfocusworld.com). (Reproduced with permission of Laser Focus World. Reproduced with permission of Block Engineering's Mini-QCL™: Miniaturized Quantum Cascade Laser OEM Module. http://blockeng.com/images/miniqcl.jpg.)

(among others) a series of semiconductor layers, corresponding to a multiple quantum well (MQW). As shown in Figure 10.11, the MQW structure is biased in a such a way that an E_l level of a well on the left coincides with an $E_r > E_l$ level in the well on the right. An electron on E_l can tunnel resonantly onto the level E_r of the next well. By emitting a photon, it goes then to the lower level, can tunnel further, and so on. This way a cascade of identical and coherent photons can be produced. Different $E_l - E_r$ pairs can be utilized by appropriate choice of the bias, so the laser can work at different wavelengths. With appropriate cooling, one can also build power lasers operating with tunable frequencies.

10.4.1 Mathematical Demonstration of Resonant Tunneling

Let us consider first the so-called one-dimensional potential trap shown in Figure 10.12, and look for solutions with $E < V_0$.

Figure 10.12 One-dimensional potential trap.

$$V(x < 0) = \infty$$
$$V(0 < x < a) = 0$$
$$V(x > a) = V_0$$

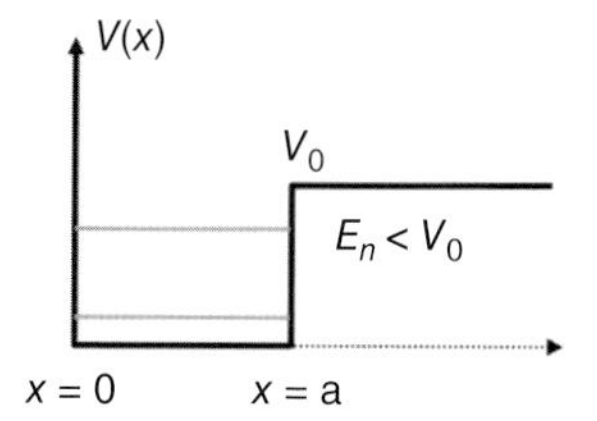

The time-independent Schrödinger equation has to be constructed in two ranges separately (for $x < 0$, φ must be zero):

$$\begin{aligned} 0 < x < a \quad &\rightarrow \quad \varphi'' = -k^2\varphi \\ x > a \quad &\rightarrow \quad \varphi'' = \frac{1}{4b^2}\varphi \end{aligned} \tag{10.17}$$

where φ'' is the second derivative of φ and, again,

$$k^2 \equiv \frac{2m}{\hbar^2}E; \quad \frac{1}{4b^2} \equiv \frac{2m}{\hbar^2}(E - V_0) \tag{10.18}$$

The solutions are then

$$\begin{aligned} &\varphi(x < 0) = 0 \\ &\varphi(0 < x < a) = A\sin(kx + B) \\ &\varphi(x > a) = Ce^{\frac{x-a}{2b}} + Be^{-\frac{x-a}{2b}} \end{aligned} \tag{10.19}$$

and the regularity condition requires that

$$\begin{aligned} &B = 0; \qquad C = 0 \\ &D = A\sin(k_n a); -D = 2k_n b_n A\cos(k_n a) \end{aligned} \tag{10.20}$$

From which we obtain the equation determining the energy eigenvalues:

$$\tan(k_n a) = -2k_n b_n$$

and with the definition of k_n:

$$tan\left(\frac{2ma}{\hbar^2}E_n\right) = -\sqrt{\frac{E_n}{V_0 - E_n}} \tag{10.21}$$

The E_n eigenvalues following from this equation do not change upon changing the potential to that in Figure 10.13.

The solutions in the four ranges are now

$$\begin{aligned} &\varphi(x < 0) = 0 \\ &\varphi(0 < x < a) = A\sin(k_n x + B) \\ &\varphi(a < x < a + d) = Ce^{\frac{x-a}{2b_n}} + Be^{-\frac{x-a}{2b_n}} \\ &\varphi(x > a + d) = Ee^{ik_n(x-a-d)} + Fe^{-ik_n(x-a-d)} \end{aligned} \tag{10.22}$$

where F and E are the amplitudes of waves propagating to the right and left, respectively. (A is the amplitude in the trap and D, G describe the electron in

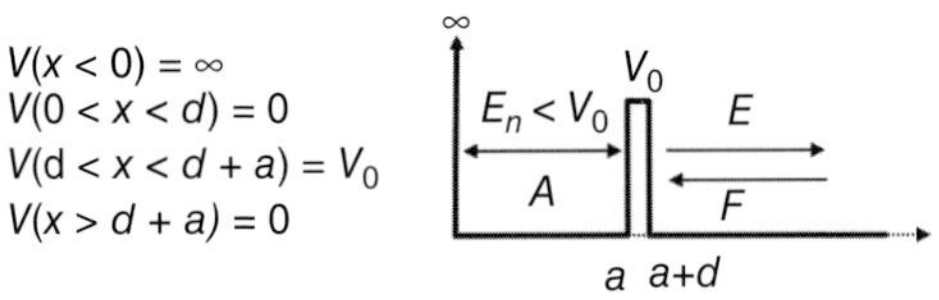

$V(x < 0) = \infty$
$V(0 < x < d) = 0$
$V(d < x < d + a) = V_0$
$V(x > d + a) = 0$

Figure 10.13 Resonant tunneling into a potential trap.

the wall.) Regularity requires that

$$\begin{aligned} 2C &= A[\sin(k_n a) + 2k_n b_n \cos(k_n a)] \\ 2D &= A[\sin(k_n a) - 2k_n b_n \cos(k_n a)] \\ 2F &= Ce^{\frac{d}{2b_n}}\left(1 - \frac{1}{2ik_n b_n}\right) + De^{-\frac{d}{2b_n}}\left(1 + \frac{1}{2ik_n b_n}\right) \end{aligned} \tag{10.23}$$

from which the trapping rate T is

$$\left|\frac{A}{F}\right|^2 = \left|\frac{1}{4}e^{\frac{d}{2b_n}}\left(1 - \frac{i}{2k_n b_n}\right)[\sin(k_n a) + 2k_n b_n \cos(k_n a)] + \frac{1}{4}e^{-\frac{d}{2b_n}}\left(1 - \frac{i}{2k_n b_n}\right)[\sin(k_n a) - 2k_n b_n \cos(k_n a)]\right|^{-2} \tag{10.24}$$

Since usually $d \gg b$ (see Section 10.4), $e^{d/2b}$ in the first term is dominant on the right-hand side, so

$$\left|\frac{A}{F}\right|^2 \propto e^{-d/b_n} \ll 1 \tag{10.25}$$

that is, the trapping rate is very low. If, however, the first term becomes zero in Eq. (10.24) because of

$$\sin(k_n a) + 2k_n b_n \cos(k_n a) = 0 \tag{10.26}$$

then

$$\left|\frac{A}{F}\right|^2 \propto e^{d/b_n} \tag{10.27}$$

becomes quite large. Equation (10.26) is, however, identical to Eq. (10.21). This means that whenever the energy of the incoming wave (of amplitude F) is equal to one of the allowed energy eigenvalues within the quantum well, the tunneling rate increases strongly. Our result is obviously valid also for tunneling out of the well and, therefore, can be applied to the double wall of Figure 10.9 as well.

Summary in Short

- Tunnel effect: Since the wave function decays exponentially into a potential wall, the probability of finding the electron on the other side of a wall with thickness $d > 2b$ is $\propto exp(-d/b)$, where the penetration depth $b \propto (V_0 - E)^{1/2}$ depends on the difference between the height of the wall and the energy of the particle.
- The tunneling current on applying a bias of $U = |\mathbf{E}|l$ is given by the Fowler–Nordheim tunnel formula, which is $\propto exp\frac{W^{3/2}}{q|\mathbf{E}|}$ (W being the work function). The tunnel effect is at work in electric plugs, in the leakage current of MOS-FETs, in the field emission of electrons, in scanning tunneling microscopy, Zener diodes, and in solid-state storage devices.

- The tunneling rate through a double wall strongly increases when the average energy of the particle is resonant with one of the allowed energy eigenvalues in the quantum well, formed by the two walls. This resonant tunnel effect is important for quantum FETs and for the quantum-cascade laser.

10.5 Questions and Exercises

Problem 10.1 An electron of energy 3 eV is approaching a potential wall of height $V_0 = 3.5$ eV and width 5 Å. The potential outside the wall is zero.

a) What is the tunneling rate for an electron ($m_e = 9.10939 \times 10^{-31}$ kg)? Compare the exponential approximation and the exact formula of Eq. (10.10).
b) Which formula would you choose for a proton ($m_p = 1.67263 \times 10^{-27}$ kg) of the same energy? What is the tunneling rate for the proton?

Problem 10.2 The band gaps of diamond and silicon are $E_g^{\text{Diam}} = 5.34$ eV and $E_g^{\text{Si}} = 1.12$ eV, respectively. Estimate the relation of the electric breakdown fields in the two materials. N.B.: comparison of the result with the experimentally observed ratio, 100/6, indicates that the effective mass of electrons (due to interaction between the wave-like electrons), is different in the two materials (see Section 5.8 and Chapter 12).

Problem 10.3 The tunnel effect is also being utilized in flash memories. These are MOS structures, with an extra, so-called floating gate (FG) in the insulating layer (Figure 10.14). Information is stored by the charge state of the FG electrode (1: charged, 0: uncharged). Writing and erasing occurs with the help of the (original) control electrode (CG). Try to explain the role of the tunnel effect in this procedure, taking into account that the thickness of the insulating layer between the semiconductor and the FG is much smaller than between the FG and the CG.

Problem 10.4 A quantum-cascade laser contains five GaAs layers of thickness 2 nm each, separated by GaAlAs barriers (six layers) of width 1 nm each, as shown idealized in Figure 10.11.

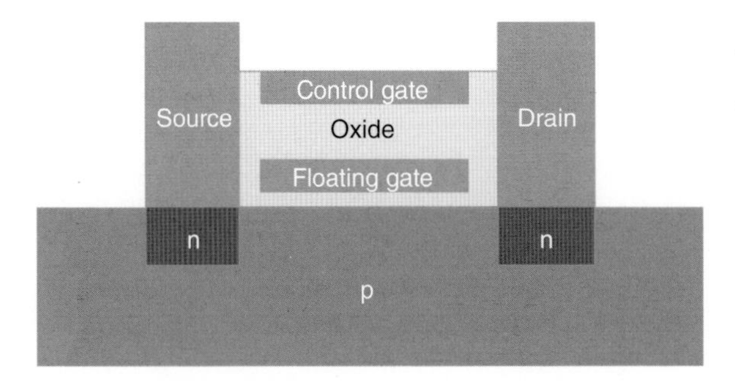

Figure 10.14 Scheme of an FG-MOS used in solid-state storage devices (flash memories).

a) Light emission occurs by the transition of the electron from the first excited state into the ground state of the wells, after it has tunneled through the separating barrier. Which wavelength will be emitted? (Use the formula for infinite quantum wells.)
b) What is the necessary bias between the two sides of the whole sandwich structure? Assume a resonant tunneling rate of 100%.

Hint: the constant field strength is given by $|\mathbf{E}| = U/l$.

11

The Hydrogen Atom. Quantum Numbers. Electron Spin

In this chapter…

We are going to consider the quantum mechanical state of the electron in a hydrogen atom, as the basis for understanding the chemical properties of the elements. In Chapter 4 on discharge lamps, we have seen that atoms can emit and absorb only discrete frequencies. In the case of hydrogen atom, these frequencies can be calculated using the Balmer formula of Eq. (4.2). In order to explain the latter, Bohr has assumed that stable orbitals of the electron in the hydrogen atom must have quantized angular momenta, $L_z = n\hbar$; see Eq. (4.5). (We have assumed here that the orbital plane is orthogonal to the z-direction, and so $\mathbf{L} = 0 \cdot \mathbf{e}_x + 0 \cdot \mathbf{e}_y + L_z \cdot \mathbf{e}_z$ and $|\mathbf{L}| = L_z$.) The hypothesis of de Broglie (Section 4.5) has traced this quantization back to the wave nature of the electron and initiated the development of quantum mechanics. After having constructed an axiomatic quantum mechanics in Chapters 6–8, we should now check what it says about the states of the electron in the hydrogen atom. The Bohr model is based on the classical concept of the point mass and on the premise that the trajectory and the momentum can both be determined at the same time. Consequently, it is assumed that all three components of the angular momentum (here 0, 0, L_z) can be determined accurately as well. However, as we have seen in Section 7.3, this runs against the uncertainty principle of quantum mechanics. As it turns out, in hindsight one must say that the Bohr model was just a lucky guess. It cannot be extended to atoms with more than one electron and, even for the hydrogen atom (while providing the correct energies), it predicts an angular momentum of $|\mathbf{L}| = L_z = \hbar$ for the ground state, in disagreement with experiment.[1] A better understanding of the atoms is, however, indispensable to understand the chemical bonds. N.B.: the covalent bond, for example, in the hydrogen molecule and also in the semiconductor materials of electronics cannot be explained by classical physics at all! In order to understand the electronic and optical properties of semiconductors and other solids, knowledge about the behavior of electrons in an atom will be necessary.

1 The motion of the electron in the atom corresponds to a circular current, which gives rise to a magnetic field, with strength depending on the angular momentum. Since the field can be measured, the angular momentum can be determined, too.

Essential Quantum Mechanics for Electrical Engineers, First Edition. Peter Deák.

11.1 Eigenstates of L_z

As mentioned in the introduction to this Chapter the stable orbits of the electron in the hydrogen atom were characterized by quantized angular momenta L_z. Let us, therefore, examine the L_z-eigenstates according to the principles of quantum mechanics. We have to solve the eigenvalue equation (Eq. (B.22)) for the operator given by Eq. (7.6):

$$\hat{L}_z \equiv \frac{\hbar}{i}\frac{\partial}{\partial\phi} \Rightarrow \frac{\hbar}{i}\frac{\partial}{\partial\phi}\cdot\varphi_m(\phi) \equiv L_{zm}\varphi_m(\phi) \tag{11.1}$$

where ϕ is the azimuth angle. This is a differential equation analogous to the eigenvalue equation of the momentum, Eq. (8.6), and has a similar solution:

$$\varphi_m(\phi) = \frac{1}{\sqrt{2\pi}} e^{i\frac{L_{zm}}{\hbar}\phi} \tag{11.2}$$

Since ϕ is restricted to the range $[0,2\pi]$, this function is square integrable and normalized by the given prefactor.[2] One should consider, however, that in a polar coordinate system ϕ and $\phi + 2\pi$ represent the same angle. To have the regularity condition satisfied, we should make sure that our solution is a single-valued function, that is,

$$\varphi_m(\phi + 2\pi) = \varphi_m(\phi) \;\rightarrow\; e^{i\frac{L_{zm}}{\hbar}(\phi+2\pi)} = e^{i\frac{L_{zm}}{\hbar}\phi} e^{i\frac{L_{zm}}{\hbar}2\pi} = e^{i\frac{L_{zm}}{\hbar}\phi} \tag{11.3}$$

This can be achieved if

$$L_{zm} = m\hbar; \quad |m| = 0, 1,2, \ldots \tag{11.4}$$

since $e^{im2\pi} = \cos(m2\pi) + i\sin(m2\pi) = 1$. Indeed, as assumed by Bohr, the eigenvalues of L_z are quantized. It is important to note that m, and with that $|L_{zm}|$, can also be zero. Now what about the other components?

11.2 Eigenstates of L^2

We have learned in Section 7.3 that $\hat{L}_z$ and $\hat{L}^2$ can have simultaneous eigenfunctions because of Eq. (7.10). Let us determine now the eigenfunctions and eigenvalues of $\hat{L}^2$. (The square root of the latter will supply $|\mathbf{L}|$.) Using Eqs (7.5)–(7.9), the $\hat{L}^2$ operator can be expressed in a polar coordinate system, that is, in terms of the azimuth angle ϕ and the polar angle ϑ. Without going into details, the result is

$$\hat{\mathbf{L}}^2 = -\hbar^2\left[\frac{\partial^2}{\partial\vartheta^2} + \cot\vartheta\frac{\partial}{\partial\vartheta} + \frac{1}{\sin^2\vartheta}\frac{\partial^2}{\partial\phi^2}\right] \tag{11.5}$$

The eigenvalue equation for this operator is then

$$-\hbar^2\left[\frac{\partial^2}{\partial\vartheta^2} + \cot\vartheta\frac{\partial}{\partial\vartheta} + \frac{1}{\sin^2\vartheta}\frac{\partial^2}{\partial\phi^2}\right] Y(\vartheta,\phi) = L^2 Y(\vartheta,\phi) \tag{11.6}$$

---

2 $\int_0^{2\pi} \frac{1}{\sqrt{2\pi}} e^{-i\frac{L_{zm}}{\hbar}\phi} \frac{1}{\sqrt{2\pi}} e^{i\frac{L_{zm}}{\hbar}\phi} d\phi = 1.$

where $Y(\vartheta, \phi)$ are the eigenfunctions of $\hat{L}^2$. Let us seek the solutions with the assumption that the eigenfunctions can be written as products of two functions, one depending only on ϑ and the other only on ϕ:

$$Y(\vartheta, \phi) = \theta(\vartheta)\varphi(\phi) \tag{11.7}$$

Substituting this into Eq. (11.6) and multiplying by $-\sin^2\vartheta/(\hbar^2\theta\varphi)$ gives

$$\frac{1}{\theta(\vartheta)}\left[\sin^2\vartheta\cdot\theta''(\vartheta) + \frac{1}{2}\sin 2\vartheta\cdot\theta'(\vartheta) + \frac{L^2\sin^2\vartheta}{\hbar^2}\theta(\vartheta)\right] = -\frac{\varphi''(\phi)}{\varphi(\phi)} \tag{11.8}$$

where the left-hand side depends only on ϑ and the right-hand side only on ϕ. (The ordinary derivatives are abbreviated here by $'$ and $''$.) As we have already discussed that in the case of Eq. (8.16), such an equation can only be satisfied if both sides are constant. For the time being, let us call this constant α. From the right-hand side, it follows then that

$$\frac{d^2\varphi(\phi)}{d\phi^2} = -\alpha\,\varphi(\phi) \quad\rightarrow\quad \varphi(\phi) = \frac{1}{\sqrt{2\pi}}e^{i\sqrt{\alpha}\,\phi} \tag{11.9}$$

In order to ensure that the solution is a single-valued function, following the argument in Section 11.1, we have to require that $\alpha = m^2$, where $|m| = 0, 1,2, \ldots$ Therefore, the regular solutions are identical to the eigenfunctions of $\hat{L}_z$:

$$\varphi_m(\phi) = \frac{1}{\sqrt{2\pi}}e^{im\phi} \tag{11.10}$$

to which the eigenvalues $L_z = m\hbar$ belong.

The left-hand side of Eq. (11.8) must also be equal to $\alpha = m^2$, and the ordinary differential equation

$$\theta''(\vartheta) + \cot\vartheta\cdot\theta'(\vartheta) + \left(\frac{L^2}{\hbar^2} - \frac{m^2}{\sin^2\vartheta}\right)\theta(\vartheta) = 0 \tag{11.11}$$

follows. This is not easy to solve. The functions satisfying it are called the *associated Legendre polynomials* in mathematics.[3] These complicated functions are, essentially, orthonormalized linear combinations of the powers of $\cos(\vartheta)$. The important point is that they are defined, besides m, by another integer l, as $\theta_l^m(\vartheta)$ and the regularity condition requires that

$$l \geq |m| \tag{11.12}$$

Using these $\theta_l^m(\vartheta)$ functions in Eq. (11.11), it can be shown that the eigenvalues of $\hat{L}^2$ are

$$L_l^2 = \hbar l(l+1) \tag{11.13}$$

3 The definition of the associated Legendre polynomials is

$$\theta_l^m(x) = \sqrt{\frac{(2l+1)}{2}\frac{(l-m)!}{(l+m)!}}(1-x^2)^{\frac{|m|}{2}}\frac{d^{|m|}P_l(x)}{dx^{|m|}}$$

where $x = \cos\vartheta$ and $P_l(x) = \sum_r^l c_r x^r$ are the so-called Legendre polynomials.

that is, the absolute value of the angular momentum is also quantized and, according to Eqs (11.12) and (11.13), the possible values thereof are determined by the value of L_z:

$$|\mathbf{L}| = \hbar\sqrt{l(l+1)}; \quad l = -m, \ldots 0, \ldots + m; \; m = \frac{L_z}{\hbar} \tag{11.14}$$

Instead of trying to analyze the $\theta_l^m(\vartheta)$ functions, let us consider the full eigenfunctions of $\hat{L}^2$, that is, $Y_l^m(\vartheta, \phi) = \theta_l^m(\vartheta)\varphi_m(\phi)$ by using Eq. (11.10):

$$Y_l^m(\vartheta, \phi) = \frac{1}{\sqrt{2\pi}}\theta_l^m(\cos\vartheta) \cdot e^{im\phi} \tag{11.15}$$

These functions are called the *spherical harmonics* in mathematics. According to Eq. (11.12), $l=0$ allows only $m=0$, while $l=1$ allows for $m=-1$, 0, +1. The corresponding spherical harmonics are relatively simple functions:

$$\begin{aligned}
Y_0^0(\vartheta, \phi) &= \frac{1}{\sqrt{4\pi}} \\
Y_1^0(\vartheta, \phi) &= \sqrt{\frac{3}{4\pi}}\cos\vartheta \\
Y_1^1(\vartheta, \phi) &= \sqrt{\frac{3}{8\pi}}\sin\vartheta \cdot e^{i\phi} \\
Y_1^{-1}(\vartheta, \phi) &= \sqrt{\frac{3}{8\pi}}\sin\vartheta \cdot e^{-i\phi}
\end{aligned} \tag{11.16}$$

and can be depicted by their isosurfaces.[4] Since the last two functions in Eq. (11.16) are complex, it is convenient to display $(Y_1^1 + Y_1^{-1})/\sqrt{2}$ and $(Y_1^1 - Y_1^{-1})/\sqrt{2}$ instead, which can be written as purely real and imaginary functions, respectively, by using the Euler formula of Eq. (B.1). Figure 11.1 shows the isosurfaces with positive values in blue and isosurfaces with negative values of the same magnitude in yellow. The top two rows depict Eq. (11.16), while the third and fourth rows show the spherical harmonics for $l = 2$ and 3.

The spherical harmonics are simultaneously eigenfunctions[5] of $\hat{L}_z$ and $\hat{L}^2$ and, as shown in Figure 11.1, they describe the allowed symmetries of the eigenstates of the angular momentum. As we discuss shortly, the electron in an atom is in such eigenstates, which resemble in no way the circular orbitals of Bohr. In chemistry, the Y_l^m functions displayed here are called s-, p-, d-, and f-orbitals. To each of them belong unique values of L_z and $|\mathbf{L}|$, while L_x and L_z have no definite values.

4 On the isosurface, the value of the function is the same everywhere. The spherical harmonics in Figure 11.1 have the same form for any value.
5 $\theta_l^m(\vartheta)$ is just a constant from the viewpoint of $\hat{L}_z$, since it does not depend on ϕ.

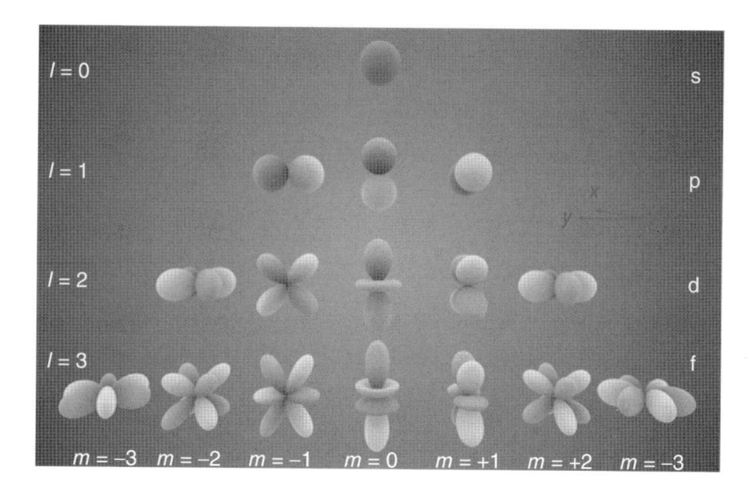

Figure 11.1 Isosurface representation of the spherical harmonics. From top to bottom: the s-, the three p-, the five d-, and the seven f-orbitals. Blue means positive and yellow negative isovalues of the same magnitude. Color online. (https://en.wikipedia.org/wiki/Spherical_harmonics Used under Creative commons license CC BY-SA 3.0 https://creativecommons.org/licenses/by-sa/3.0/.)

For the functions in Eq. (11.16),

$$\begin{aligned} |\mathbf{L}| &= 0\,; \quad L_z = 0 \;\Rightarrow\; Y_0^0 \equiv s \\ |\mathbf{L}| &= \hbar\sqrt{2}\,; \quad L_z = 0 \;\Rightarrow\; Y_1^0 \equiv p_z \\ |\mathbf{L}| &= \hbar\sqrt{2}\,; \quad L_z = +\hbar \;\Rightarrow\; Y_1^1 \equiv p_x \\ |\mathbf{L}| &= \hbar\sqrt{2}\,; \quad L_z = -\hbar \;\Rightarrow\; Y_1^{-1} \equiv p_y \end{aligned} \tag{11.17}$$

It can be seen that for $|\mathbf{L}| \neq 0$, $|\mathbf{L}| \neq |L_z|$, that is, in contrast to the assumption of Bohr, L_x and L_y cannot be zero.

11.3 Energy Eigenstates of an Electron in the Hydrogen Atom

The electron in the hydrogen atom is subjected to the Coulomb force of the proton in the nucleus. Having the origin of the coordinate system on the proton, the corresponding potential energy

$$V(r, \vartheta, \phi) = V(r) = -\frac{q^2}{4\pi\varepsilon_0 r} \tag{11.18}$$

(where $q = -e$ is the elementary charge and ε_0 is the dielectric constant), depends only on the distance r of the electron from the nucleus. Therefore, it is reasonable to use a polar coordinate system for the time-independent Schrödinger equation. The kinetic energy operator

$$\hat{T} = -\frac{\hbar^2}{2m}\Delta = -\frac{\hbar^2}{2m}\left(\frac{\partial^2}{\partial x^2} + \frac{\partial^2}{\partial y^2} + \frac{\partial^2}{\partial z^2}\right) \tag{11.19}$$

contains the Laplace operator (second derivative), which can (with the help of Footnote 5 in Section 7.3, or directly from some handbook of mathematics similar to that of I. N. Bronshtein *et al.*) be written in polar coordinates as

$$\Delta_{r,\vartheta,\phi} = \frac{\partial^2}{\partial r^2} + \frac{2}{r}\frac{\partial}{\partial r} + \frac{1}{r^2}\underbrace{\left(\frac{\partial^2}{\partial \vartheta^2} + \cot\vartheta \frac{\partial}{\partial \vartheta} + \frac{1}{\sin^2\vartheta}\frac{\partial^2}{\partial \phi^2}\right)}_{-\hat{L}^2/\hbar}$$

$$= \frac{\partial^2}{\partial r^2} + \frac{2}{r}\frac{\partial}{\partial r} - \frac{1}{r^2}\frac{\hat{L}^2}{\hbar^2} \tag{11.20}$$

Comparison with Eq. (11.5) shows that the Laplace operator in polar coordinates contains the $\hat{L}^2$ operator. This indicates that angular momentum will characterize the state of the electron, as expected in a central force field.[6] The time-independent Schrödinger equation for the electron in the hydrogen atom is then

$$\left[-\frac{\hbar^2}{2m}\left(\frac{\partial^2}{\partial r^2} + \frac{2}{r}\frac{\partial}{\partial r} - \frac{1}{r^2}\frac{\hat{L}^2}{\hbar^2}\right) - \frac{q^2}{4\pi\varepsilon_0}\frac{1}{r}\right]\varphi(r,\vartheta,\phi) = E\varphi(r,\vartheta,\phi) \tag{11.21}$$

Since we expect the solutions to be also eigenstates of the angular momentum (i.e., of $\hat{L}_z$ and $\hat{L}^2$), let us seek them in the form of

$$\varphi(r,\vartheta,\phi) = F(r)Y(\vartheta,\phi) \tag{11.22}$$

Substitution into Eq. (11.21) provides[7] the equation for the radial function $F(r)$:

$$F'' + \frac{2}{r}F' + \left[\frac{2m}{\hbar^2}(E - V) - \frac{l(l+1)}{r^2}\right]F = 0 \tag{11.23}$$

This is another ordinary differential equation of second order, which is not easy to solve. The solutions can be built from the so-called *Laguerre polynomials*. Besides l, these $F_n^l(r)$ functions[8] are characterized by yet another integer n, and the regularity condition requires that

$$n > l \tag{11.24}$$

---

6 In classical physics we know that the angular momentum is constant in a central force field.
7 After multiplication by $-(2m/\hbar^2)(r^2/FY)$, Eq. (11.21) can be brought to the form

$$r^2 \cdot \frac{F''(r)}{F(r)} + 2r \cdot \frac{F'(r)}{F(r)} + \frac{2m}{\hbar^2}[E - V(r)]r^2 = \frac{\hat{L}^2 Y(\vartheta,\varphi)}{\hbar^2 Y(\vartheta,\varphi)}$$

where, again, the two sides depend on different variables and therefore must be constant. Because of Eq. (11.3), this condition is satisfied for the right-hand side with $l \cdot (l+1)$. Therefore the left-hand side must also be equal with this constant, and Eq. (11.23) follows.
8 The solution of Eq. (11.23) is

$$F_n^l(r) = -\left[\frac{(n-l-1)!}{2n}\right]^{1/2} \cdot \left[\frac{2}{(n+l)!na_B}\right]^{3/2} \cdot \left(\frac{2r}{na_B}\right)^l \cdot p_{n+l}^{2l+1}\left(\frac{2r}{na_B}\right) e^{-\frac{r}{na_B}}$$

Where

$$p_{n+l}^{2l+1}(x) = \frac{d^{2l+1}}{dx^{2l+1}}\left\{e^x\left[\frac{d^{l+n}}{dx^{l+n}}(x^{l+n}e^{-x})\right]\right\}$$

are the Laguerre polynomials defined by the integers n and l, whereas $x = 2r/r_n$ and $a_B = 4\pi\varepsilon_0(\hbar^2/mq^2)$.

Equation (11.23) with the radial functions $F_n^l(r)$ supplies the quantized energy eigenvalues as

$$E_n = -\frac{me^4}{8\varepsilon_0^2 h^2}\frac{1}{n^2} = \frac{-13.605}{n^2}(\text{eV}) \tag{11.25}$$

which are *independent* of l. This result happens to be identical to Eq. (4.9), that is, with Bohr's result for the energy. However, Bohr had to exclude $n=0$ (i.e., a zero radius) arbitrarily, while from Eqs (11.24) and (11.13) $n=1, 2,\ldots$ follows automatically. (This is also in agreement with the uncertainty principle, which requires a zero point energy; cf. Section 9.5.) The energy eigenfunctions

$$\varphi_{n,l,m}(r, \vartheta, \phi) = F_n^l(r)Y_l^m(\vartheta, \phi) \tag{11.26}$$

are simultaneously eigenstates for the absolute value of the angular momentum with the quantized eigenvalues

$$|L_l| = \hbar\sqrt{l(l+1)} \tag{11.27}$$

and for the z-component of the angular momentum with the quantized eigenvalues

$$L_{zm} = m\hbar \tag{11.28}$$

The integers n, l, and m are called the *principal, angular momentum,* and *magnetic quantum numbers*, respectively.[9] They define the possible eigenstates of Eq. (11.26), and they determine the corresponding eigenvalues of the energy, the modulus, as well as one component of the angular momentum, according to Eqs (11.25), (11.27), and (11.28), respectively. The conditions in Eqs (11.4) and (11.24) mean that several (degenerate) eigenfunctions of different angular momenta can belong to any given energy E_n:

<table>
<tr><td>n</td><td>1</td><td colspan="4">2</td><td colspan="9">3</td></tr>
<tr><td>l</td><td>0</td><td>0</td><td colspan="3">1</td><td>0</td><td colspan="3">1</td><td colspan="5">2</td></tr>
<tr><td>m</td><td>0</td><td>0</td><td>−1</td><td>0</td><td>1</td><td>0</td><td>−1</td><td>0</td><td>1</td><td>−2</td><td>−1</td><td>0</td><td>1</td><td>2</td></tr>
<tr><td></td><td>1s</td><td>2s</td><td colspan="3">2p</td><td>3s</td><td colspan="3">3p</td><td colspan="5">3d</td></tr>
</table>

(11.29)

The orbitals are usually denoted by the principal (energy) quantum number and by the chemical name of the spherical harmonics, which describes the angular momentum eigenstate. The orbitals, belonging to a given principal quantum number, are together called the *electron shell.*

Here again, depicting the electronic states is more informative than looking at the general mathematical formula. The radial functions of s-orbitals ($l=0$) are

9 The angular momentum (or orbital angular momentum) quantum number is often called the azimuthal quantum number. The magnetic momentum of the electron depends on L_z, and that is why m is called the magnetic quantum number.

relatively simple:

$$n = 1 \quad l = 0 \quad F_1^0(r) = \frac{2}{\sqrt{a_B^3}} e^{-r/a_B}$$

$$n = 2 \quad l = 0 \quad F_2^0(r) = \frac{2}{\sqrt{(2a_B)^3}} e^{-r/a_B} \left(1 - \frac{r}{2a_B}\right) e^{-r/2a_B}$$

$$n = 2 \quad l = 1 \quad F_3^0(r) = \frac{3}{3\sqrt{(3a_B)^3}} \left[3 - 6\frac{r}{3a_B} + 2\left(\frac{r}{3a_B}\right)^2\right] e^{-r/3a_B} \qquad (11.30)$$

where $a_B = 4\pi\varepsilon_0(\hbar^2/mq^2)$. As can be seen in the upper panels of Figure 11.2, these functions have $n-1$ nodes. The sign changes at the nodes, so if we multiply the s-orbitals, shown in Figure 11.1, by these radial functions, we obtain the picture of the eigenfunctions $\varphi_{n,0,0}(r, \vartheta, \phi)$ for $n = 1, 2, 3$ in the lower panels of Figure 11.2. For p-orbitals ($l = 1$), the $F_n^1(r)$ functions have $n-2$ nodes but the spherical harmonics $Y_1^m(\vartheta, \phi)$ themselves have a node (see Figure 11.1), resulting in the 2p- and 3p-states shown in Figure 11.3. Figure 11.4 provides an overview of the $n \geq 2$ states in cross section.

These pictures clearly show that the amplitude of the wave function is not distributed along a circular orbit (not to mention about localization in a single geometrical point) and, in the case of the s-orbitals, the amplitude is even highest in the center. One should consider, however, that the radial functions of Eq. (11.30) contain

$$a_B = 4\pi\varepsilon_0 \frac{\hbar^2}{mq^2} \qquad (11.31)$$

in the exponent, which is the first Bohr radius appearing in Eq. (4.8)! This means that the Bohr radii may have some meaning after all in quantum mechanics too. Bohr has assumed a circular orbit for the electron, so let us ask the question: what

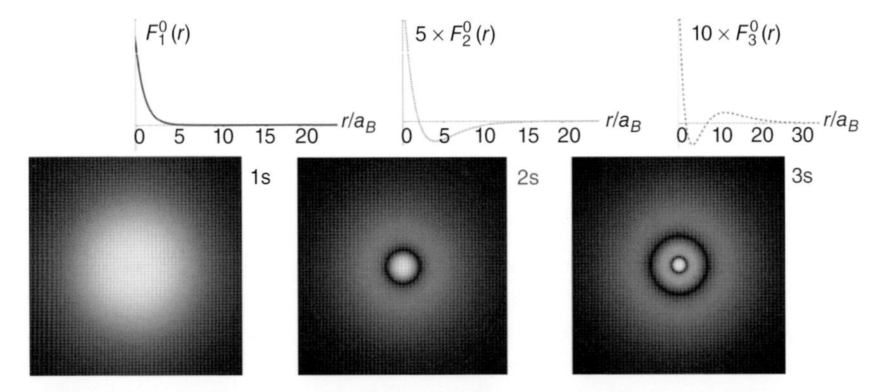

Figure 11.2 The 1s-, 2s-, and 3s-orbitals $F_n^0(r)Y_0^0(\vartheta, \phi)$ of the electron in a hydrogen atom. Positive values are shown in yellow, negatives in blue, and the color intensity indicates the variation of the magnitude, as given by the radial function $F_n^0(r)$, shown above the pictures. Color online. Reproduced with permission of Wolfgang Christian (http://www.compadre.org/pqp/).

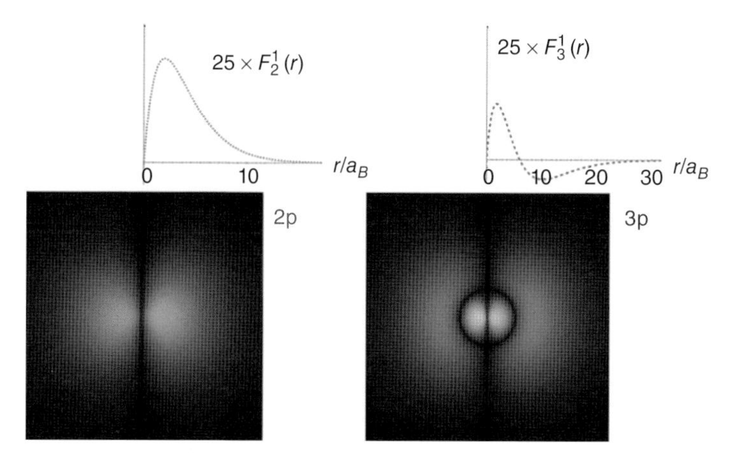

Figure 11.3 The $2p_x$, and $3p_x$-orbitals $F_n^1(r)Y_1^1(\vartheta, \phi)$ of the electron in a hydrogen atom. Positive values are shown in yellow, negatives in blue, and the color intensity indicates the variation of the magnitude, as given by the radial function $F_n^0(r)$, shown above the pictures. Color online. Reproduced with permission of Wolfgang Christian (http://www.compadre.org/pqp/).

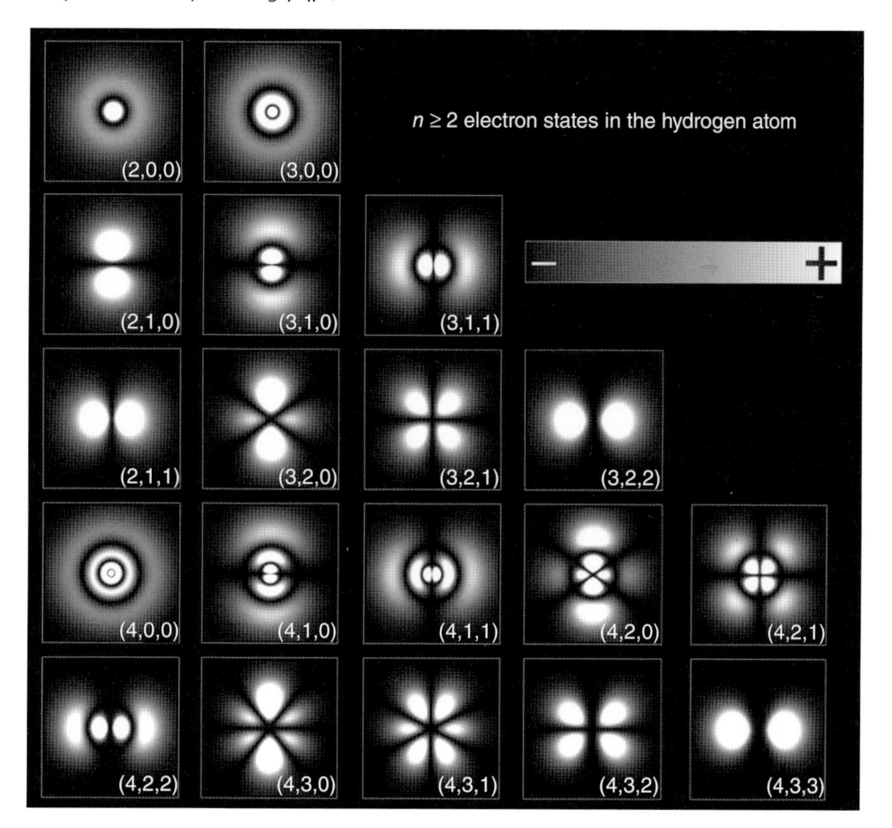

Figure 11.4 Overview of the electron orbitals in the hydrogen atom for $n \geq 2$. The square of the wave function (i.e., the probability density of finding the electron in the space around the nucleus) is depicted. The variation of the magnitude is color coded from yellow = high to blue = small values. Color online. (https://commons.wikimedia.org/wiki/File:Hydrogen_Density_Plots.png Used under Creative commons license CC BY-SA 3.0 https://creativecommons.org/licenses/by-sa/3.0/.)

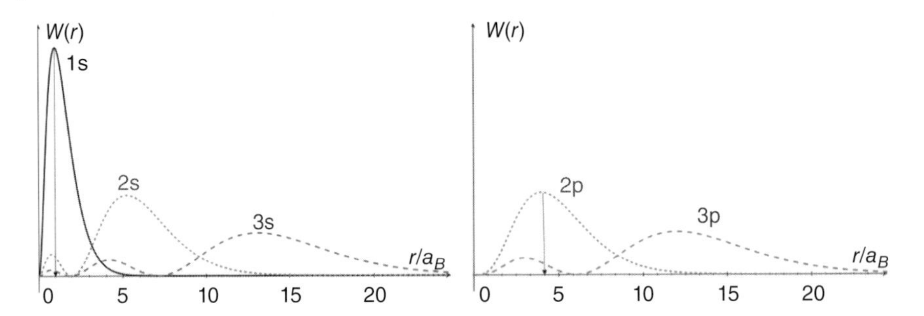

Figure 11.5 The radial probability distribution of the electron in various states in the hydrogen atom.

is the probability of finding the electron in a distance r from the nucleus? The probability of finding the electron at any *one* point is given by $|F_n^l(r)|^2$. However, there are $4\pi r^2$ points with a distance r from the nucleus, so the answer to our question is given by the *radial probability distribution*

$$W_n^l(r) = 4\pi r^2 |F_n^l(r)|^2 \tag{11.32}$$

As can be seen in Figure 11.5, this function can have several maxima, and it can be shown that, the orbitals $l = n - 1$ (and only those) have their main maximum at

$$r_n = n^2 a_B \tag{11.33}$$

which corresponds to the Bohr radii of Eq. (4.8). The Bohr radii provide the most likely distance of the electron from the nucleus – but for the $l = n - 1$ states only. The Bohr model is, therefore, a genial but very limited approximation of the real situation in the hydrogen atom.

11.4 Angular Momentum of the Electrons. The Spin

The angular momentum of the electron is connected to its magnetic moment (a quantity that describes the reaction of the electron to a magnetic field) and can, therefore, be measured. The measurements have shown that electrons, independent of their orbital, in fact even outside an atom, always have an own angular momentum, called *spin*. Despite the name, the spin has nothing to do with rotation.[10] It is a characteristic of the particle, just as its mass or charge. The spin (*own angular momentum*) is a vector with the same relation between the components as the *orbital angular momentum*: only one component and the modulus can be determined accurately at the same time. For the spin vector **S** of an electron, these are always

$$|\mathbf{S}| = \frac{\sqrt{3}}{2}\hbar; \quad S_z = \pm\frac{1}{2}\hbar \tag{11.34}$$

10 Historically, the electron was thought of earlier as a rotating sphere on a circular orbit around the nucleus, similar to the earth around the sun. The spin was attributed to the rotation in this model.

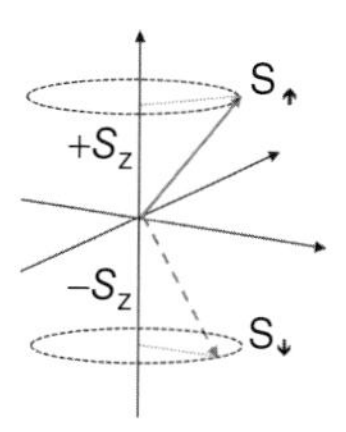

Figure 11.6 Possible orientations of the electron spin. (The end point of the vector must be on the upper or the lower circle.)

This means that the spin vector can only have two possible orientations: with an angle of 54.736° to the positive or to the negative z-axis (Figure 11.6). These are called the *spin-up* and *spin-down* states of the electron.

In the hydrogen atom, an electron in any orbital can have a spin-up or a spin-down state. In order to describe the state unambiguously, the set of quantum numbers have to be amended by the *spin quantum number*, $s = \pm 1$ (corresponding to up and down spin):

$$\begin{aligned} &n = 1,\ 2,\ 3,\ 4, \ldots \\ &l = 0, 1, 2, \ldots, n \\ &m = -l, \ldots,\ 0, \ldots, +l \\ &s = -1, +1 \end{aligned} \tag{11.35}$$

The total angular momentum of the electron is given by the sum of the orbital angular momentum and the spin, $\mathbf{J} = \mathbf{L} + \mathbf{S}$ and, so, $J_z = L_z + S_z$.

According to Eqs (11.25)–(11.28), an electron in the ground state ($n = 1$) of the hydrogen atom has the energy $E_1 = -13.605$ eV and an orbital angular momentum of $|\mathbf{L}| = L_z = 0$ ($l = 0$). This means that the total angular momentum of the ground state comes solely from the spin. Quantum mechanics predicts, therefore, $|\mathbf{J}| = |\mathbf{S}| = (\sqrt{3}/2)\hbar$ and $J_z = S_z = \pm(\hbar/2)$ for the ground state, which has been confirmed by measurements. In contrast, the Bohr model assumes $|\mathbf{L}| = L_z = \hbar$.

The spin state should somehow be shown by the wave function too. Since the spin is independent of the spatial variables, it can be described by a multiplying function, which depends on the spin quantum number:

$$\phi_{n,l,m,s}(r, \vartheta, \varphi) = \phi_{n,l,m}(r, \vartheta, \varphi) \cdot \sigma(s) = F_n^l(r) Y_l^m(\vartheta, \varphi) \cdot \sigma_s \tag{11.36}$$

Summary in Short

- In the Coulomb potential of the atomic nucleus, $V(r) = -(Ze^2)/(4\pi\varepsilon_0 r)$ (where Z is the number of protons), the (stationary) energy eigenstates of *one* electron are simultaneously eigenstates of $\hat{L}^2$ and $\hat{L}_z$.
- The eigenvalues of E, $|\mathbf{L}|$, and L_z are quantized:

$$\begin{aligned} &E_n = -\frac{m_e e^4}{8\varepsilon_0^2 h^2} \frac{Z^2}{n^2}; \quad n = 1,2, \ldots \\ &L_l = \hbar\sqrt{l(l+1)}; \quad l = 0,1, \ldots, n \\ &L_{zm} = \hbar m; \quad m = -l, \ldots, 0, \ldots, +l \end{aligned}$$

  where the quantum numbers n, l, and m are prescribed by the regularity condition.

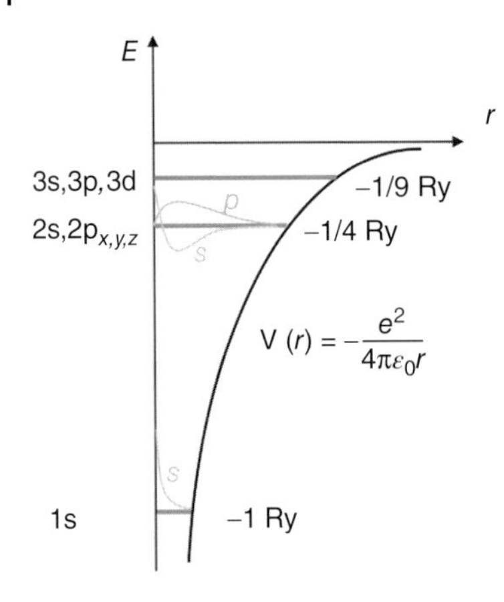

Figure 11.7 The discrete energy levels of *one* electron in the field of *Z* protons. The radial parts of the orbitals are depicted using the corresponding levels as *r*-axis. Color online.

- The energy of the hydrogen atom ($Z = 1$) in the ground state is 1 Rydberg = −13.605 eV, and the orbital angular momentum is $|\mathbf{L}_0| = 0$. This shows that the electron is not a point mass moving along a circular orbit. Higher energy eigenstates can have different angular momenta, given by the spherical harmonics,

$$Y_{l=0}^{m=0} \equiv s; \quad Y_{l=1}^{m=0} \equiv p_z, \quad Y_{l=1}^{m=1} \equiv p_x, \quad Y_{l=1}^{m=-1} \equiv p_y; \quad \text{etc.}$$

- The atomic orbitals are energy and angular momentum eigenfunctions, constructed as the product of a spherical harmonics $Y_l^m(\vartheta, \varphi)$ and a radial function $F_n^l(r)$. The latter have nodes, increasing in number with energy ($n - 1$ for s-orbitals, $n - 2$ for p-orbitals), as expected from a bound state (see Figure 11.7).
- The radial probability distribution $W_n^l(r)$ of the atomic orbitals (or, rather, of the "electron clouds") can have several maxima. The main maximum of the orbital with $l = n - 1$ is at the nth Bohr radius, $r_n = 4\pi\varepsilon_0\hbar^2/(Ze^2m)n^2$.
- Every elementary particle has a characteristic spin **S** (own angular momentum). For the electron, $|\mathbf{S}| = (\sqrt{3}/2)\hbar$ and $S_z = \pm(1/2)\hbar$. The state of the electron in the atom is given by the
  main or energy quantum number n
  angular or azimuthal quantum number l
  magnetic quantum number m
  spin quantum number s (which is either +1 or −1).

11.5 Questions and Exercises

Problem 11.1 The stationary states of the electron in the hydrogen atom are characterized by four quantum numbers.

a) Which observables are provided by these quantum numbers and how?
b) What are the allowed values of the quantum numbers?

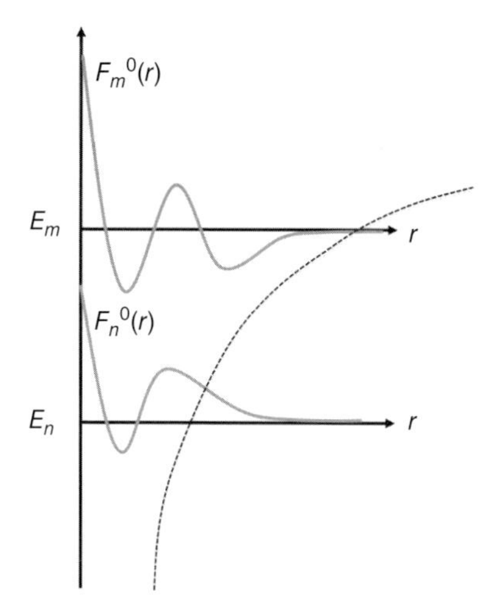

Figure 11.8 Two electron states in the hydrogen atom, depicted by their radial part using the energy levels in the Coulomb potential as abscissa.

Problem 11.2 Consider the electron states in the hydrogen atom with energy −0.8503 eV.

a) What is the value of the corresponding main quantum number?
b) How many states can have this energy? Set up a table for the possible quantum numbers!
c) What is the difference between these states of the same energy?
d) Provide a general formula for the polar angle (to the z-axis) of the angular momentum! Can this angle be zero?
e) Assuming all possible states (including spin) to be occupied by an electron, calculate the sum of the z-components of the total angular momentum!

Problem 11.3 Which wavelength is radiated at the electron transition from orbital m, depicted in Figure 11.8, to the orbital n? The ionization energy of the hydrogen atom is −13.605 eV.

Problem 11.4 What does quantum mechanics say about the atomic orbitals and how does that differ from the Bohr model?

12

Quantum Mechanics of Many-Body Systems (Postulates 6 and 7). The Chemical Properties of Atoms. Quantum Information Processing

In this chapter…
We consider the consequences of the wave nature of electrons in systems with more than one electron. Our five postulates so far have all been set up with a single electron in mind, with the rest of the world being merely the source of the potential energy $V(\mathbf{r},t)$, which determines the state of the observed electron. For example, the hydrogen atom, described in the previous chapter, consists of the proton in the nucleus and of the electron. We have, however, considered the proton only as the "environment" of the electron, providing the Coulomb potential. We could do that because the proton is much bigger than the electron, so – to a good approximation – it can be considered as a classical, fixed point mass (with zero uncertainty of its position). Going from hydrogen to bigger atoms (not to speak about molecules and solids), however, we will have to consider the interaction of the electrons, that is, the coexistence of several (many) identical quantum mechanical particles. For such cases, we will have to add two further postulates to the five we already have. These will lead to the understanding of the differing chemical properties of atoms. We also encounter the entanglement of electron states, which is being used for quantum encrypting and is likely to be exploited for quantum computing.

12.1 The Wave Function of a System of Identical Particles

With several electrons in a system, their wave nature becomes particularly important. As for classical waves, we can expect superposition, after which the individual electrons will not be distinguishable anymore. *The whole system must be described by a single, common wave function. The absolute square of this function should provide the probability density of finding the first electron at* $\mathbf{r}_1$*, the second at* $\mathbf{r}_2$*, and so on. The wave function of a system of N electrons has, therefore, N space variables, and should be square-integrable* ($\int |\Psi(\mathbf{r}_1, \mathbf{r}_2, \ldots, \mathbf{r}_N, t)|^2 d\mathbf{r}_1 d\mathbf{r}_2 \cdots d\mathbf{r}_N = 1$). Since the individual electrons are not distinguishable anymore, the measurable quantity, that is, the absolute square of

Essential Quantum Mechanics for Electrical Engineers, First Edition. Peter Deák.

the wave function, must not change if two space variables are exchanged. From this follows that

$$|\Psi(\dots,\mathbf{r}_i,\dots,\mathbf{r}_j,\dots,\mathbf{t})|^2 = |\Psi(\dots,\mathbf{r}_j,\dots,\mathbf{r}_i,\dots,\mathbf{t})|^2$$
$$\Rightarrow \begin{cases} \Psi(\dots,\mathbf{r}_i,\dots,\mathbf{r}_j,\dots,\mathbf{t}) = +\Psi(\dots,\mathbf{r}_j,\dots,\mathbf{r}_i,\dots,\mathbf{t}) \\ \Psi(\dots,\mathbf{r}_i,\dots,\mathbf{r}_j,\dots,\mathbf{t}) = -\Psi(\dots,\mathbf{r}_j,\dots,\mathbf{r}_i,\dots,\mathbf{t}) \end{cases} \tag{12.1}$$

that is, the wave function must be symmetric or antisymmetric to the exchange of two coordinates.[1] With this consideration, we must add another postulate to quantum mechanics:

Postulate 6

The state of a system of N particles can be described by a single, regular wave function $\Psi = \Psi(\mathbf{r}_1,\dots,\mathbf{r}_N,t)$, which must be either symmetric or antisymmetric to the exchange of the coordinates of two *identical* particles.

N.B.: The symmetry/antisymmetry requirement also applies for the exchange of spin coordinates.

12.2 The Pauli Principle

In addition to Postulate 6, *Wolfgang Pauli* has found that a correct description of a system of electrons[2] is only possible with establishing yet another:

Postulate 7

The wave function of a system of electrons can only be antisymmetric for the exchange of coordinates:

$$\Psi(\dots\mathbf{r}_i,\dots,\mathbf{r}_j,\dots t) = -\Psi(\dots\mathbf{r}_j,\dots,\mathbf{r}_i,\dots t) \tag{12.2}$$

In order to understand the consequences of this mathematical requirement, let us consider a helium atom with two electrons. Since the nucleus has two protons, the first electron is moving in the potential field $V_1 = -(2q^2)/(4\pi\varepsilon_0 r_1)$ and the second in $V_2 = -(2q^2)/(4\pi\varepsilon_0 r_2)$, where $q = -e$ is the elementary charge and ε_0 is the dielectric constant. The origin of the coordinate system is the point-mass-like nucleus for both electrons. The Hamilton operator of this *two-electron system* should contain, besides the multiplicative operators of the aforementioned potential energies, the kinetic energy operators for both electrons

$$\hat{T}_i = -\frac{\hbar^2}{2m}\left(\frac{\partial^2}{\partial x_i^2} + \frac{\partial^2}{\partial y_i^2} + \frac{\partial^2}{\partial z_i^2}\right) = -\frac{\hbar^2}{2m}\Delta_i; \quad i = 1,2 \tag{12.3}$$

---

1 As an example, the function $\exp[ik(x_1 + x_2)]$ is symmetric while, say, e^{ikx_2}/x_1 is neither symmetric nor antisymmetric.

2 And all other particles with spin according to Eq. (11.34). Such particles are called *fermions*.

and the potential energy operator due to their interaction:

$$\hat{V}_{12} = \frac{q^2}{4\pi\varepsilon_0|\mathbf{r}_1 - \mathbf{r}_2|}. \tag{12.4}$$

Therefore,

$$\hat{H} = \underbrace{(\hat{T}_1 + \hat{V}_1)}_{\hat{H}_1} + \underbrace{(\hat{T}_2 + \hat{V}_2)}_{\hat{H}_2} + \hat{V}_{12} = \underbrace{(\hat{H}_1 + \hat{H}_2)}_{\hat{H}^0} + \hat{V}_{12} = \hat{H}^0 + \hat{V}_{12} \tag{12.5}$$

As can be seen, this is a sum of two operators, the first ($\hat{H}^0$) describing two independent electrons moving in the potential field of the same nucleus and the second ($\hat{V}_{12}$) describing the interaction between the electrons. A time-independent Schrödinger equation provides the stationary state of the two-electron system. Was it not for the interaction part of the operator, we could easily find the solution for

$$\hat{H}^0\Phi^0(\mathbf{r}_1, \mathbf{r}_2) = E^0\Phi^0(\mathbf{r}_1, \mathbf{r}_2) \tag{12.6}$$

Since both V_1 and V_2 differ only in the factor $Z = 2$ from the case of the hydrogen atom in Eq. (11.18), the functions in Eq. (11.36) will satisfy the eigenvalue equations

$$\begin{aligned} \hat{H}_1\varphi_{n,l,m,s}(\mathbf{r}_1) &= \varepsilon_{1n}\varphi_{n,l,m,s}(\mathbf{r}_1) \\ \hat{H}_2\varphi_{n,l,m,s}(\mathbf{r}_2) &= \varepsilon_{1n}\varphi_{n,l,m,s}(\mathbf{r}_2) \end{aligned} \tag{12.7}$$

with identical eigenvalues (cf. the summary at the end of the previous chapter):

$$\varepsilon_{1n} = \varepsilon_{2n} = -\frac{me^4}{8\varepsilon_0^2h^2}\frac{4}{n^2} \tag{12.8}$$

(N.B.: The exponent of the radial part of the orbitals will also be larger than for hydrogen by a factor $Z = 2$, that is, the $F_n^l(r)$ functions will decay faster.) From the single-particle wave functions, we can construct a special solution for Eq. (12.6):

$$\Phi^0_{\text{spec}}(1, 2) = \varphi_1(1)\varphi_2(2) \tag{12.9}$$

where the space variables have been abbreviated as $(\mathbf{r}_1) \rightarrow (1)$ and $(\mathbf{r}_2) \rightarrow (2)$, and the indices of φ denote two possible combinations of the quantum numbers n, l, m, and s. Substituting Eq. (12.9) into Eq. (12.6) and considering that operators with index 1 (or 2) affect only the arguments 1 (or 2), it can then be seen that Eq. (12.9) is indeed an eigenfunction

$$\begin{aligned} (\hat{H}_1 + \hat{H}_2)\varphi_1(1)\varphi_2(2) &= (\hat{H}_1\varphi_1)\varphi_2 + \varphi_1(\hat{H}_2\varphi_2) \\ &= (\varepsilon_{1n}\varphi_1)\varphi_2 + \varphi_1(\varepsilon_{2n}\varphi_2) = (\varepsilon_{1n} + \varepsilon_{2n})\varphi_1\varphi_2 \end{aligned} \tag{12.10}$$

with the eigenvalue $E^0 = \varepsilon_{1n} + \varepsilon_{2n}$. This result can be generalized. In a system of N independent (noninteracting) particles, the energy eigenfunction of the system can be written as the product of the single-particle wave functions, and the total energy is the sum of the single-particle energies. So, if $\hat{H}^0 = \sum_i \hat{H}_i$ and

$\hat{H}_i \varphi_i(\mathbf{r}_i) = \varepsilon_i \varphi_i(\mathbf{r}_i)$, then a special solution of the eigenvalue problem is

$$\Phi^0_{\text{spec}}(\mathbf{r}_1, \ldots, \mathbf{r}_N) = \prod_{i=1}^{N} \varphi_i(\mathbf{r}_i); \quad E^0_{\text{tot}} = \sum_{i=1}^{N} \varepsilon_i \tag{12.11}$$

However, in our two-electron system, Eq. (12.9) is only one special solution, and others such as $\varphi_1(1)\varphi_1(2)$, $\varphi_2(1)\varphi_1(2)$, and $\varphi_2(1)\varphi_2(2)$ are also possible. (In other words, one can have both electrons in either of the two orbitals or the first electron in φ_1 and the second in φ_2, or vice versa.) The general solution of Eq. (12.6) is a linear combination of the special solutions:

$$\Phi^0(1,2) = c_{11}\varphi_1(1)\varphi_1(2) + c_{12}\varphi_1(1)\varphi_2(2) + c_{21}\varphi_2(1)\varphi_1(2) + c_{22}\varphi_2(1)\varphi_2(2) \tag{12.12}$$

and the absolute square of the coefficients can be considered as the probability of the corresponding special solutions. However, the wave function of a many-electron system must comply with Postulate 7, that is, fulfill Eq. (12.2):

$$\Phi^0(1,2) = -\Phi^0(2,1) \tag{12.13}$$

In the case of Eq. (12.12) that is only possible if $c_{11} = 0,\ c_{22} = 0$, and $C_{12} = C_{21}$.[3] This means that *the probability of having both electrons in the same n, l, m, s state is zero!* That is why the Pauli principle is often formulated as follows: **two electrons cannot have the same state** or **any energy and angular momentum eigenstate can be occupied by up to two electrons with opposite spins.** It should be noted though that, in coming to this conclusion, we have assumed that the independent singe-particle states survive in the two-electron system, which is only possible if we neglect the interaction term in Eq. (12.5). The generally valid formulation of the Pauli principle is given by Postulate 7.

12.3 Independent Electron Approximation (One-Electron Approximation)

If the interaction of the electrons could be neglected, the stationary state of the two-electron system in the helium atom, which complies with Postulates 6 and 7, would be

$$\Phi^0(12) = \frac{1}{\sqrt{2}}[\varphi_1(1)\varphi_2(2) - \varphi_2(1)\varphi_1(2)] \tag{12.14}$$

This can be generalized for an arbitrary number N of identical particles with the help of a so-called *Slater determinant*, replacing Eq. (12.11), which does not fulfill the Pauli principle, by

$$\Phi^0(1\ldots N) = \frac{1}{\sqrt{N!}} \begin{vmatrix} \varphi_1(1) & \varphi_2(1) & \cdots & \varphi_N(1) \\ \varphi_1(2) & \varphi_2(2) & \cdots & \varphi_N(2) \\ \vdots & \vdots & \ddots & \vdots \\ \varphi_1(N) & \varphi_2(N) & \cdots & \varphi_N(N) \end{vmatrix} \tag{12.15}$$

3 The value of the latter is determined to be $\pm 1/\sqrt{2}$ by the normalization of the wave function.

where the indices $i = 1, 2, \ldots, N$ correspond to possible combinations of the quantum numbers n, l, m, and s, that is, to single-particle atomic functions $\varphi_{n,l,m,s}$ similar to those of Eq. (11.25).

The stationary state of the real system (with Coulomb interaction between the electrons) can be obtained from the time-independent Schrödinger equation with the full Hamilton operator of Eq. (12.5):

$$\hat{H}(1 \ldots N)\Phi_i(1 \ldots N) = E_1\Phi_i(1 \ldots N) \tag{12.16}$$

The total energy in the eigenstate Φ_i can be expressed formally from this equation (after scalar multiplication by Φ_i) as

$$E_i = \frac{\langle \Phi_i | \hat{H}\Phi_i \rangle}{\langle \Phi_i | \Phi_i \rangle} \tag{12.17}$$

It can be proven that using an arbitrary function $\tilde{\Phi}$ (which is not an eigenfunction of $\hat{H}$) on the right-hand side of this equation, the resulting energy

$$\hat{E}[\tilde{\Phi}] = \frac{\langle \tilde{\Phi} | \hat{H}\tilde{\Phi} \rangle}{\langle \tilde{\Phi} | \tilde{\Phi} \rangle} \tag{12.18}$$

will always be higher than the E_0 energy of the ground state Φ_0. This allows an approximation of the true Φ_0 and E_0: one should search for the function $\tilde{\Phi}$ which makes the energy $\tilde{E}$ minimal. This is procedure is called the *variational principle.* The question is: what should be the starting guess for $\tilde{\Phi}$ which allows for the minimization process? The many-body state of an "electron-soup," in which the single electrons lose their individuality, is difficult to imagine and a connection to the classical theory of electric conduction, based on the flow of point-mass-like charge carriers (see Ohm's law in Eq. (A.12)) appears to be hardly possible. We can, however, make use of the fact that in the construction of $\tilde{\Phi}$ only the end result counts – the picture we put behind it does not! If the electrons were independent (without Coulomb interaction), we could construct the many-body wave function as in Eq. (12.15). *So let us define fictitious particles with the charge and spin of an electron in some (as yet unknown) one-particle states φ_i, which – despite the charge – do not interact with each other.* We call these independent particles *one-electrons.* We can then construct $\tilde{\Phi}$ as in Eq. (12.15), and the variational principle allows us to determine the one-electron wave functions φ_i, which make $\tilde{E}$ minimal, that is, $\tilde{\Phi}$ to be the closest to Φ_0. One can expect that the resulting φ_i functions, describing nominally independent fictitious particles, will nevertheless ensure that the one-electrons behave as if they were influenced by the Coulomb field of the others. In other words, we "encode" the Coulomb interaction, at least approximately, into the properties of the one-electrons one by one. As a rough analogy, one can think of a beer crate as the real system, divided into 20 compartments by a grid. One could, however, just as well construct a fully functional beer crate by gluing 20 small cases together. Having 20 compartments *together* in the beer crate is a property of the "many-body system," which cannot be achieved by the 20 individual "single-particles," unless modifying their properties to account for the Coulomb interaction, that is, applying a glue on the surface of each small case.

Without going into details, the minimization of Eq. (12.18), with $\tilde{\Phi}$ constructed as in Eq. (12.15), leads to the following equation for each one-electron wave function:

$$\left(-\frac{\hbar^2}{2m}\Delta_i + V[\varphi_1, \ldots, \varphi_N]\right)\varphi_i = \varepsilon_i\varphi_i \tag{12.19}$$

where the potential energy operator consists of three terms:

$$\hat{V}(\varphi_1 \ldots \varphi_N) \equiv \hat{V}_{\text{ext}} + \hat{V}_c(\varphi_1 \ldots \varphi_N) + \hat{V}_X(\varphi_1 \ldots \varphi_N) \tag{12.20}$$

here V_{ext} is the external potential energy due to the presence of other particles outside the many-electron system (e.g., the nucleus in the helium atom), and

$$\hat{V}_c(\varphi_1 \ldots \varphi_N) \equiv \frac{q^2}{4\pi\varepsilon_0}\sum_j \int \frac{\varphi_j^*(\mathbf{r}')\varphi_j(\mathbf{r}')}{|\mathbf{r}' - \mathbf{r}|} d\mathrm{r}' \cdot \varphi_i(r)$$

$$\hat{V}_X(\varphi_1 \ldots \varphi_N) \equiv -\frac{1}{2}\frac{q^2}{4\pi\varepsilon_0}\sum_j \int \frac{e^2\varphi_j^*(\mathbf{r}')\varphi_i(\mathbf{r}')}{|\mathbf{r}' - \mathbf{r}|} d\mathrm{r}' \cdot \varphi_j(r) \tag{12.21}$$

The first expression contains $\rho(\mathbf{r}') = q[\varphi_j^*(\mathbf{r}')\varphi_j(\mathbf{r}')]$, which can be considered as the "spatial charge distribution" of a one-electron in state φ_j. (The absolute square of a wave function gives the probability density of finding the particle in space.) We can recognize the Coulomb interaction of this charge distribution with a one-electron (charge q) in state φ_i. That is why this term is called the *Coulomb operator*. The summation means that the Coulomb interaction of a one-electron in state φ_i with all other one-electrons is taken into account (at least in an averaged way). The second expression V_X is called the eXchange operator, since the indices *i* and *j* are exchanged with respect to V_C. Essentially, it describes a Coulomb interaction as well and appears as a consequence of the Pauli principle (i.e., due to applying Eq. (12.15) instead of Eq. (12.11)). The important thing is, however, that the variational principle leads to the same Eq. (12.19) for every one-electron state φ_i, and that is in the same form as a single-particle Schrödinger equation. Actually, the potential energy operator depends, through Eq. (12.21), on the one-electron states that have to be determined,[4] so Eq. (12.19) can only be solved by iteration.[5] The potential, $V(\varphi_1 \ldots \varphi_2)$, constructed from the final solution for the wave functions is called the SCF (self-consistent field) potential.

The SCF procedure can only be carried out on a computer. Nonetheless, we are able to introduce independent particles having the charge of an electron. When one speaks about "electrons" in a semiconductor, that is, about individual charge carriers, the flow of which determines the current (as in classical physics), these fictitious one-electrons are actually meant. The electron–electron interaction is encoded into the properties of φ_i by Eq. (12.19). This will have, for example, the

4 Therefore, Eq. (12.19) is called a *pseudo* Schrödinger equation.

5 For example, one neglects V_C and V_X first and determines a set of starting eigenfunctions $\{\phi_i^0\}$ from Eq. (12.19), which can then be used to construct these potentials. Then, Eq. (12.19) is solved again and a new set $\{\phi_i^1\}$ is obtained. The procedure must be continued till the total energy (sum of the one-electron energies ε_i, see Eq. (12.11)) changes between two consecutive steps less than the required accuracy.

consequence that the mass of the one-electrons may differ from the real electron mass. From the dispersion relation of free (real) electrons in Eq. (8.26), it follows that

$$m = \left[\frac{1}{\hbar^2} \frac{d^2E(k)}{dk^2} \right]^{-1}$$

The $E(k)$ relation of the one-electrons in a solid is a complicated (non-parabolic) function and that can lead to effective masses, which are different from that of a real electron. These masses vary according to the wave number, that is, with the speed ($p = mv = \hat{h}k$). This can lead to the change of the mass with external bias, which explains the deviation from Ohm's law, shown in Figure 5.9. Actually, in some materials, the mass of the charge carriers can become zero. This happens, for example, in graphene, allowing for extreme high speed of these so-called Dirac electrons.

12.4 Atoms with Several Electrons

Solving Eq. (12.19) by an SCF procedure for an atom with several electrons leads to φ_i single-particle states, which are actually very similar to the orbitals given by Eq. (11.36) for the hydrogen atom, and can also be characterized by the quantum numbers n, l, m, and s. However, the one-electron energies will be dependent – in contrast to Eq. (11.25) – not only on n but also on l. While in hydrogen, for example, the 3s-, 3p-, and 3d-states have the same energy, this degeneracy is split in atoms with more than one electron.

Figure 12.1 shows how many of the one-electrons can have the same $E_{n,l}$ energy in such atoms. (For any given l, the states with $m = -l, \ldots, 0, \ldots, l$ will still have the same energy and in each of them the spin quantum number can be $s = \pm 1$.) In the ground state of the atom with N electrons, the lowest energy levels will be occupied according to this scheme, obeying the Pauli principle. As can be seen in Figure 12.1, the 4s states will be occupied before the 3d states, because of splitting.[6]

12.5 The Chemical Properties of Atoms

The discipline chemistry contains our knowledge about how the atoms can build molecules (and solids). The chemical bond between atoms cannot be described simply by the Coulomb interaction of point charges: it arises dominantly due to the quantum mechanical interaction of the electron clouds. This is depicted schematically in Figure 12.2.

The Coulomb potentials of the two nuclei give rise to a potential wall between the two atoms. The one-electron states penetrate the wall, as shown by the radial part of the s-orbitals for each $E_{n,o}$ energy. One-electrons of the two atoms can only interact with each other if their wave functions overlap. On the one hand,

6 Similarly, 5s precedes 4d, 6s precedes 4f and 5d, and 7s precedes 5f and 6d.

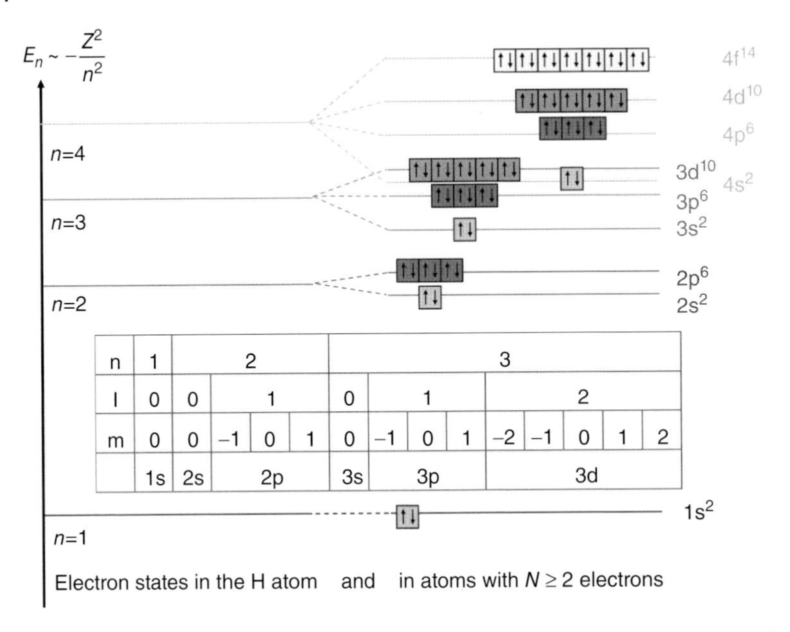

Figure 12.1 Splitting and occupation of the energy levels in an atom with many electrons. Color online.

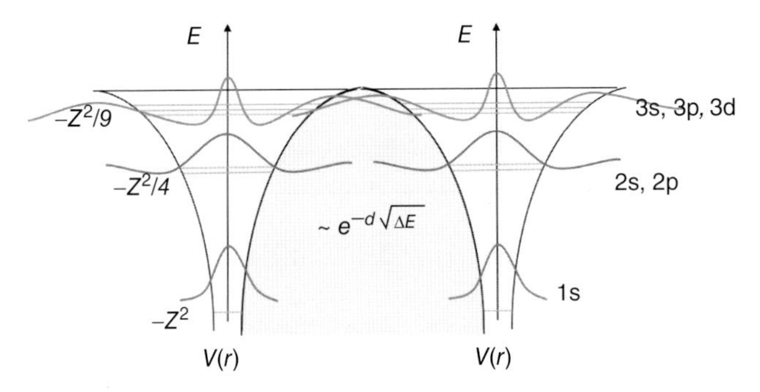

Figure 12.2 Interaction between the electrons of two atoms. For details, see the text. Color online.

in the case of approximately equal energies, the tunneling probability increases resonantly (cf. Section 10.4). On the other hand, according to Eqs. (10.12–13), it decreases exponentially with the width of the wall and with the square root of the difference between the one-electron energy and the top of the potential wall. Therefore, the interaction is strongest between the states in the highest lying electron shell, while all other interactions are usually negligibly small.[7] From the viewpoint of the chemical bond, the only relevant electron shell is the highest one, which is occupied by electrons. This shell is called the *valence shell*, and the one-electrons occupying it are the *valence electrons*. The one-electrons on the

7 N.B.: For example, $Z = 6$ for carbon and $Z = 14$ in silicon, so the lower lying energy levels are very far from the top of the wall.

deeper lying shells are called *core electrons*. The chemical properties of the atom, that is, how it can enter a chemical bond, are determined by the occupation of the valence shell. As we have seen in Section 11.3, the s-, p-, d-, and f-orbitals are similar in symmetry and orientation for all energy quantum numbers n. Therefore, we can expect atoms with the same number of valence electrons in s-, p-, d-, and so on, orbitals to have similar chemical properties.

12.6 The Periodic System of Elements

With an increasing number of electrons in the atom, the one-electron states of increasing energy are getting occupied one after the other: after 1s and 2s, the 2p states, then 3s, 3p, and (as shown in Figure 12.1) first 4s, then 3d, and so on.[8] Figure 12.3 shows identical occupations of the s-, p-, d-, and f-orbitals in the valence shells with different n values, set under each other. Atoms of the chemical elements (chemically different atoms) can be arranged according to this scheme. Chemically different atoms have different atomic numbers Z, that is, different number of protons in the nucleus. In the neutral charge state of the atom, the number of electrons equals that of protons. For example, lithium (Li) has 3, sodium (Na) 11 electrons. The three electrons of Li occupy the 1s and 2s orbitals:

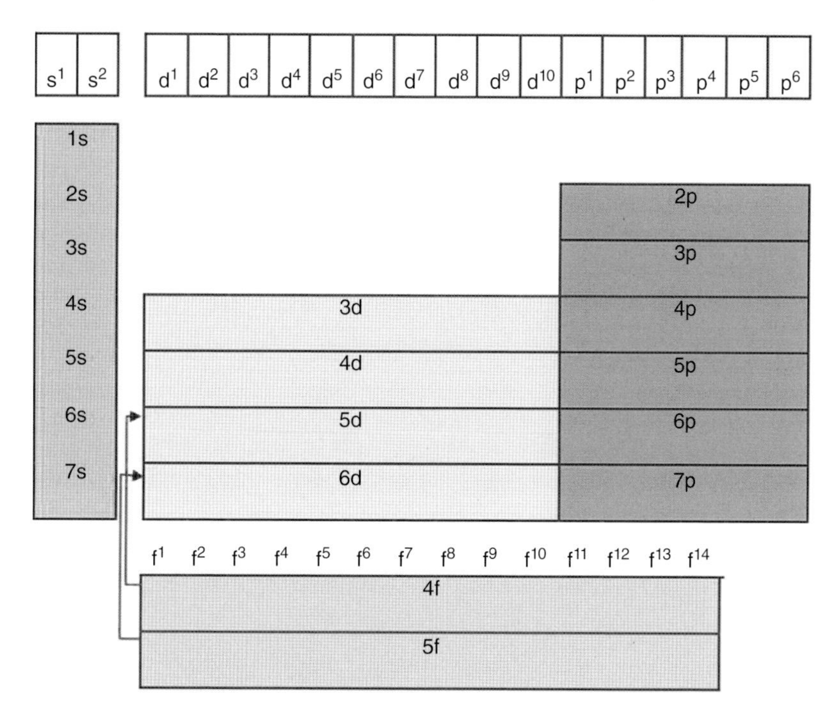

Figure 12.3 Occupation sequence of the atomic orbitals with periodically repeated s-, p-, d-, and f-character for the valence shell.

8 Actually, s means one, p three, d five, and f seven states of the same energy, and each of them can be filled with spin up and down.

Group	1	2		3	4	5	6	7	8	9	10	11	12	13	14	15	16	17	18
Period																			
1	1 H																		2 He
2	3 Li	4 Be												5 B	6 C	7 N	8 O	9 F	10 Ne
3	11 Na	12 Mg												13 Al	14 Si	15 P	16 S	17 Cl	18 Ar
4	19 K	20 Ca		21 Sc	22 Ti	23 V	24 Cr	25 Mn	26 Fe	27 Co	28 Ni	29 Cu	30 Zn	31 Ga	32 Ge	33 As	34 Se	35 Br	36 Kr
5	37 Rb	38 Sr		39 Y	40 Zr	41 Nb	42 Mo	43 Tc	44 Ru	45 Rh	46 Pd	47 Ag	48 Cd	49 In	50 Sn	51 Sb	52 Te	53 I	54 Xe
6	55 Cs	56 Ba	*	71 Lu	72 Hf	73 Ta	74 W	75 Re	76 Os	77 Ir	78 Pt	79 Au	80 Hg	81 Tl	82 Pb	83 Bi	84 Po	85 At	86 Rn
7	87 Fr	88 Ra	**	103 Lr	104 Rf	105 Db	106 Sg	107 Bh	108 Hs	109 Mt	110 Ds	111 Rg	112 Uub	113 Uut	114 Uuq	115 Uup	116 Uuh	117 Uus	118 Uuo

*Lanthanides	*	57 La	58 Ce	59 Pr	60 Nd	61 Pm	62 Sm	63 Eu	64 Gd	65 Tb	66 Dy	67 Ho	68 Er	69 Tm	70 Yb
**Actinides	**	89 Ac	90 Th	91 Pa	92 U	93 Np	94 Pu	95 Am	96 Cm	97 Bk	98 Cf	99 Es	100 Fm	101 Md	102 No

Figure 12.4 Periodic system of the elements.

$1s^2 2s^1$. In this case, $1s^2$ belongs to the core and the valence shell is $2s^1$. In Na, the occupation is $1s^2 2s^2 2p^6 3s^1$. The shells $n = 1$ and 2 belong to the core and the valence shell is $3s^1$. Therefore, Na is placed in the box "$3s^1$," below Li, which is in the box "$2s^1$." In a similar manner comes potassium (K) with the atomic number 19 ($1s^2 2s^2 2p^6 3s^2 3p^6 4s^1$) in the box "$4s^1$" under Na. And really, the atoms Li, Na, and K in the first column of this scheme are chemically very similar.

The periodic repetition of the chemical properties in the elements ordered according to atomic weight was noticed already by *Mendeleev* in the nineteenth century. The reason for the periodicity in the *modern periodic system of the elements* (see Figure 12.4) can be explained by quantum mechanics only.

12.7 Significance of the Superposition States for the Future of Electronics and Informatics

The independent electron (or one-electron) approximation, introduced in Section 12.4, is the basis for understanding the currents in integrated circuits (ICs) and, therefore, the devices of solid state electronics (diodes, amplifiers, logic gates), but also those of optoelectronics (light emitting diodes, photodiodes) and photovoltaics (solar cells).[9] New developments in these areas occurred thus far by continuous miniaturization, which has led to the well-known *Moore*'s law. As shown in Figure 12.5, the characteristic size of the devices shrunk and their number on the same chip has grown more or less exponentially for over 40 years. As mentioned in Chapter 10 in relation to tunneling, this development has its limits, and in recent years a slowdown can be observed. Much more importantly, it seems that, since 2010, further miniaturization has not led to a cost decrease. Apparently, new ways have to be found for further development.

9 The route from this chapter to the explanation of all these devices is planned to be shown in a next book: *Essential Semiconductor Physics for Engineers*.

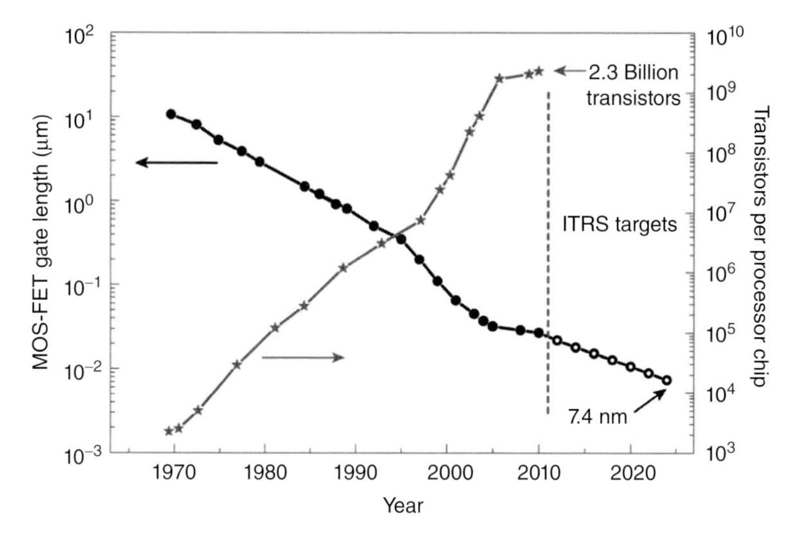

Figure 12.5 International road map for semiconductors (Moore's law). (Schwierz 2010 [1], Reproduced with permission of Nature Publishing.)

Possibly the most daring way toward a new type of electronics is quantum information processing. *Today's MOS-based logic gates are processing and producing bits, having the unambiguous values of "0" or "1."* The new information theory, which moves away from the *Boolean* (yes or no) logic in the algorithms, cannot be described in this book[10] but the hardware realization is based on quantum mechanical superposition states.

The so-called "qubits" (quantum bits) *are in a "whether–nor" state:* $\alpha \cdot |0\rangle + \beta \cdot |1\rangle$, *that is, in a superposition of two discrete quantum mechanical states,* similar to Eq. (12.14), which describes the superposition of two possibilities: electron 1 with up and electron 2 with down spin corresponding to $|0\rangle$, or electron 1 with down and electron 2 with up spin, corresponding to $|1\rangle$. An attempt to measure the spin of one of the electrons has a 50% chance of resulting in spin up or spin down. However, no matter which is the actual outcome, the other electron will have the opposite spin by a 100% probability. This is called *entanglement* of the spins. If one of the electrons could be set by some procedure into the spin-up state, the other would assume *instantaneously* (without any delay) the spin-down state. Unfortunately, in the case of an atomic two-electron state this possibility cannot be utilized, for the electrons are "too close" for influencing them separately.

One can, however, achieve entanglement also between photons. In some so-called optically nonlinear crystals, a single photon of energy $h\nu$ can be converted into two photons of energy $h\nu/2$ each, in such a way that the latter remain entangled. Measuring, for example, the polarization state of one of

10 A few recommended introductions are as follows:

E. G. Rieffel, W. H. Polak, Quantum Computing: A Gentle Introduction
M. A. Nielsen, I. L. Chuang, Quantum Computing and Quantum Information
N. D. Mermin, Quantum Computer Science.

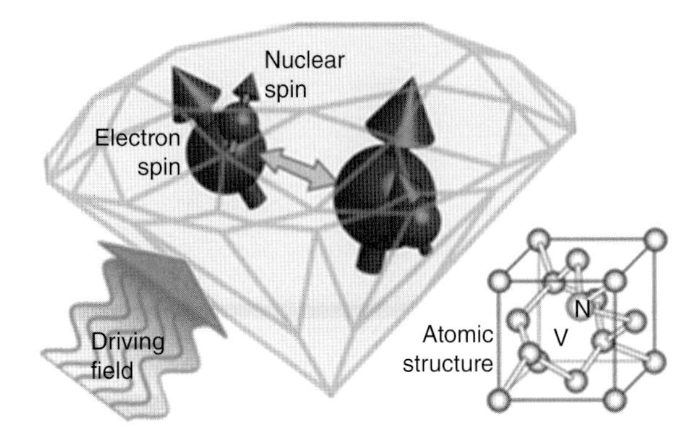

Figure 12.6 Entanglement between the spins of two nitrogen + vacancy defects by a photon in a diamond crystal. (Benjamin and Smith [2]. Color online. http://physics.aps.org/articles/v4/78?referer=rss&ref=nf. Used Under License CC By-SA 3.0, https://creativecommons.org/ licenses/by-sa/3.0/.)

them will instantaneously set the polarization state of the other, even if the two photons are far from each other.[11]

Entanglement between two particles can also be achieved by the mediation of another particle. In superconductors, electron pairs are formed due to their interaction with the lattice vibrations (phonons). The so-called *Cooper* pairs have a total spin of zero and are not subject to the Pauli principle. Consequently, any number of them can be in the energy ground state and can move without scattering, that is, carry a current without resistivity. The two electrons of a pair can be quite far from each other but their spin is entangled: when one is "up" the other must be "down." It is possible to set the spin of one electron and, thereby, set the spin of the distant other instantaneously. The existing experimental confirmation for this refutes Einstein's aversion from "spooky interaction at a distance." The coupling of the entangled Cooper pairs are, however, very weak and that is why the temperature must be very low. This would make application in quantum information processing quite expensive.

The most promising possibility so far is the entanglement of the spins of defect atoms in a crystal, by the mediation of a photon. An example for this is a nitrogen atom substituting a carbon atom next to a vacancy in the diamond crystal (Figure 12.6).

The total spin of the electrons and of the nitrogen nucleus has two well-defined arrangements, both of which can be set following a protocol of consecutive photon absorption and emission events. The spin of two such defects can be entangled by a photon. In 2013, this was demonstrated already at a distance of 3 m between the defects (in separate diamond crystals) [3]! The entanglement between the defects allows to utilize them as qubits in the algorithms of a quantum computer. Presently, experiments are ongoing to build architectures for quantum simulators, which will allow upscaling [4].

If and when quantum computing becomes a market reality, one cannot yet predict. One thing is sure however: the possibilities offered by the quasi-classical picture of the electric current, based on the one-electron approximation, will

11 Unfortunately, this cannot be used for information transfer above the speed of light, for measuring either polarization state of the first photon will have 50% probability. Still, if, for example, the entanglement is disturbed by an observer, the statistics at the receiving site will change and the act of the eavesdropping can be detected. This can be used in encrypting.

soon be exhausted and new developments in information processing and transmission will require a detailed understanding of quantum mechanics. Electrical engineers obtaining their diploma nowadays will definitely be challenged by this paradigm shift during their career.

Summary in Short

- An N-electron system is described by a single Ψ wave function with N spatial (and spin) variables. Exchange of two variables must make the wave function change sign. This is the general mathematical formulation of the requirements that identical particles are indistinguishable and two electrons cannot occupy the same quantum state (Pauli principle).
- It is possible to replace a system of interacting electrons by a system of independent, fictitious *one-electrons* in such a way that Ψ remains approximately the same. The condition of equivalence (the variational principle) leads to a pseudo Schrödinger equation, $[\hat{T} + \hat{V}(\phi_j)]\phi_i = \varepsilon_i\phi_i$, which can be used to determine the one-electron wave functions $\{\varphi_i\}$ by iteration. The potential of this equation "encodes" the interaction of the real electrons into the φ_i wave function of the one-electrons, and

$$\Psi = \Phi e^{i\frac{E}{\hbar}t} \leftarrow E = \sum_i \varepsilon_i; \quad \Phi = \frac{1}{\sqrt{N!}} \begin{vmatrix} \phi_1(1) & \phi_2(1) & \cdots \\ \phi_1(2) & \phi_2(2) & \vdots \\ \vdots & \ddots & \cdots \end{vmatrix}$$

  The fictitious one-electrons retain their individuality but can have properties different from that of a real electron, because of the quantum mechanical interactions hidden in their wave function.
- Atoms beyond hydrogen are many-electron systems and must be treated like that. The pseudo Schrödinger equation leads to one-electron states, which are very similar to the orbitals of the hydrogen atom; however, the energies of the states with the same energy quantum number, but with different angular momentum quantum numbers will be different. This splitting of the degeneracy with respect to hydrogen, together with the Pauli principle, explains the periodic system of elements, considering the fact that the chemical bond is formed primarily by the valence electrons.
- The many-body states of identical particles are superposition states, in which the properties of the individual particles are entangled, that is, setting the state of one of them sets at once the other. Entanglement between two particles can be mediated by a third particle. Such entangled states can be used as qubits, that is, as units for building a quantum computer.

12.8 Questions and Exercises

Problem 12.1 The z-component of the photon spin is $\pm\hbar$. What kind of a symmetry property must a multi-photon wave function have?

Problem 12.2 What is the energy of the ground state of eight noninteracting electrons in a one-dimensional, infinite quantum well of width L (with $V = 0$ at the bottom of the well)? According to your expectation, how would the energy change if interactions between the electrons are not neglected?

Problem 12.3 The valence electrons of a semiconductor can be considered, to a first approximation, as being trapped in a very wide quantum well. If applying the one-electron approximation, how are the interactions between the electrons taken into account? The fictitious independent one-electrons can be considered to be the charge carriers in the electric current. What consequences can the hidden quantum mechanical interactions have on their properties?

Problem 12.4 Try to explain why the one-electron energies form a band in metals (cf. Figure 3.2)!

Problem 12.5 One cannot generate really random numbers by using traditional logic gates, while quantum computers can very well do that. Try to explain that considering what can be expected when measuring (reading) a qubit that is in the superposition state $(1/\sqrt{2})(|0\rangle + |1\rangle)$!

References

1 Schwierz, F. (2010) *Nat. Nanotechnol.*, **5**, 487–496.
2 Benjamin, S.C. and Smith, J.M. (2011) *Physics*, **4**, 78.
3 Bernien, H. *et al.* (2013) *Nature*, **497**, 86.
4 Cai, J. *et al.* (2013) *Nat. Phys.*, **9**, 168.

A

Important Formulas of Classical Physics

A.1 Basic Concepts

A.1.1 The Point Mass

Macroscopic bodies consist of elementary particles (according to our present knowledge, of quarks and leptons), which appear to be negligibly small in size. Therefore, *idealized* elementary *particles* can be regarded as *point masses*, that is, geometrical points with a mass.

A.1.2 Frame of Reference

a) The position of a point mass in *space* can only be given with respect to selected other objects, setting up a *coordinate system* (origin and directions).
b) The motion of a point mass can be followed in *time* only with respect to a selected event ($t = 0$) and by comparing it to a periodic event (clock rate).

A.1.3 The Path

The position of a point mass (particle) as a function of time in a given frame of reference can be determined, in principle, accurately. This function is called the path or orbit of the particle and can be characterized by its derivatives:

$$\text{1D}: \quad \underset{\text{Path}}{x = x(t)} \rightarrow \underset{\text{Velocity}}{\frac{dx(t)}{dt} = v(t)} \rightarrow \underset{\text{Acceleration}}{\frac{d^2x(t)}{dt^2} = a(t)} \tag{A.1}$$

$$\text{3D}: \quad \mathbf{r} = \mathbf{r}(t) \rightarrow \frac{d\mathbf{r}(t)}{dt} = \mathbf{v}(t) \rightarrow \frac{d^2\mathbf{r}(t)}{dt^2} = \mathbf{a}(t) \tag{A.2}$$

A.1.4 Kinematics[1]

Uniform Motion with Constant Acceleration

1) along a straight line, see Figure A.1a, with $a = \text{constant}$, $x(0) = x_0, y(0) = y_0, v_x(0) = v_0$.

1 (Mathematical Framework for Determining the Path from the Knowledge of the Acceleration)

Essential Quantum Mechanics for Electrical Engineers, First Edition. Peter Deák.

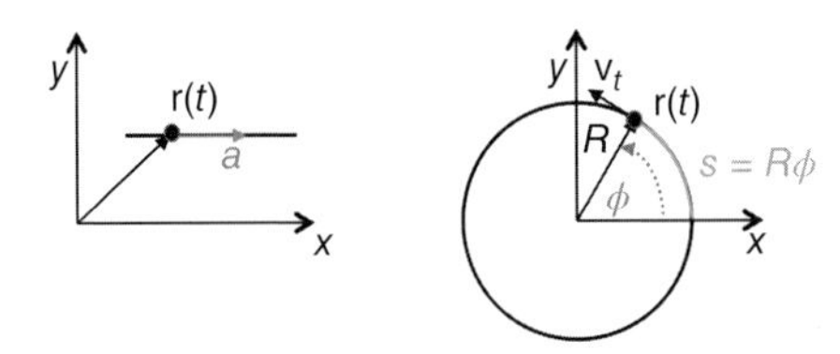

Figure A.1 Description of the motion along a straight line and along a circle. Color online.

$$\ddot{y}(t) = 0 \Leftarrow \dot{y}(t) = 0 \Leftarrow y(t) = y_0$$

$$\ddot{x}(t) = a \Rightarrow \dot{x}(t) = \int \ddot{x}(t)dt = at + v_0$$

$$\Rightarrow x(t) = \int \dot{x}(t)dt = \frac{a}{2}t^2 + v_0 t + x_0 \tag{A.3}$$

2) along a circle, see Figure A.1b, with R = constant, v_t = constant (but not the vector $\mathbf{v}_t$!), $\beta \equiv \ddot{\phi}$ = constant, and $\phi(0) = \phi_0$.

$$\omega \equiv \dot{\phi} = \frac{v_t}{R}, \quad (\boldsymbol{\omega} \times \mathbf{R} = \mathbf{v}_t); \quad a_{cp} = \frac{v_t^2}{R} = R\omega^2 \tag{A.4}$$

$$\ddot{r}(t) = 0 \Leftarrow \dot{r}(t) = 0 \Leftarrow r(t) = R$$

$$\ddot{\phi}(t) = \beta = \text{constant} \Rightarrow \dot{\phi}(t) \equiv \omega(t)$$

$$= \beta t + \omega_0 \Rightarrow \phi(t) = \frac{1}{2}\beta t^2 + \omega_0 t + \phi_0 \tag{A.5}$$

If ω = constant, this corresponds to harmonic vibrations in the Cartesian coordinates:

$$x = R\cos\phi = R\cos(\omega t + \phi_0)$$

$$y = R\sin\phi = R\sin(\omega t + \phi_0)$$

with the angular velocity/frequency of $\omega = 2\pi\nu = 2\pi\frac{1}{T}$, where ν is the frequency and T is the period.

A.2 Newton's Axioms

Definition: if a point mass, not subjected to the action of any force, moves uniformly along a straight line with respect to a given frame of reference, that frame of reference is called an *inertial system* (IS).

(I) *Axiom (condition of validity)*: The Newtonian equation of motion is valid only if the chosen frame of reference is an IS:
N.B. if the frame of reference is not an IS, the equation of motion has to be corrected by the so-called inertial forces. For example, if the origin of the chosen coordinate system has an acceleration $\mathbf{A}(t)$, and the axes rotate with the angular velocity $\boldsymbol{\Omega}(t)$, with respect to an IS, the necessary correction is $-m\mathbf{A} - m\boldsymbol{\Omega} \times (\boldsymbol{\Omega} \times \mathbf{r}') - 2m(\boldsymbol{\Omega} \times \mathbf{v}') - m\dot{\boldsymbol{\Omega}} \times \mathbf{r}'$.

(II) *Axiom (equation of motion)*: the acceleration of a body is determined by the force ($\mathbf{F}$) and depends on its inertia (m), so that $m\frac{d^2\mathbf{r}(t)}{dt^2} = \mathbf{F}(\mathbf{r}, t)$
N.B.: Knowing $\mathbf{F}(\mathbf{r},t)$, the path, $\mathbf{r}(t)$, can be determined from the initial conditions $\mathbf{r}(0) = \mathbf{r}_0, \mathbf{v}(0) = \mathbf{v}_0$.

(III) *Axiom (principle of interaction)*: Forces can only arise due to the symmetric interaction of two bodies. If body 1 acts on body 2, then the latter also acts on the former and $\mathbf{F}_{12} = -\mathbf{F}_{21}$.

(IV) *Axiom (principle of superposition)*: Forces act independently and can, therefore, all be substituted with their vector sum $\mathbf{F} = \sum_i \mathbf{F}_i$.
N.B.: Using this axiom, a given force can be substituted also by the vector sum of conveniently chosen fictitious forces. (For example, one can split the acting force into two fictitious components parallel to the Cartesian axes.)

A.3 Conservation Laws

The effect of the environment (other bodies) on a point mass is expressed by the force $\mathbf{F}$ but can also be characterized by the torque $\mathbf{M} \equiv \mathbf{r} \times \mathbf{F}$ and by the work $W \equiv \int \mathbf{F} d\mathbf{r}$ done by the force, quantities depending on both the force and the path. From the equation of motion, it follows that $\mathbf{F}$, $\mathbf{M}$, and W determine the change of the following *dynamic observables*:

Dynamic observable	Definition		Principle		Conservation	
(Linear) *momentum*	$\mathbf{p} \equiv m\mathbf{v}$	$\rightarrow$	$\frac{d\mathbf{p}}{dt} = \mathbf{F}$	$\rightarrow$	if $\mathbf{F} = 0$, $\mathbf{p} =$ constant	(A.6)
Angular momentum	$\mathbf{L} = \mathbf{r} \times \mathbf{p}$	$\rightarrow$	$\frac{d\mathbf{L}}{dt} = \mathbf{M}$	$\rightarrow$	if $\mathbf{M} = \mathbf{r} \times \mathbf{F} = 0$, $\mathbf{L} =$ constant	(A7)
Kinetic energy	$T \equiv \mathbf{p}^2/(2m)$	$\rightarrow$	$\Delta T = W$	$\rightarrow$	if $\mathbf{F} = -\nabla V, E \equiv T + V =$ constant	(A8)

The $\mathbf{p}$, $\mathbf{L}$, and T are conserved (are constant in time) if the determining quantities, $\mathbf{F}$, $\mathbf{M}$, and W are zero. (N.B.: The latter two can be zero even if $\mathbf{F}$ is not.) If the force can be written as the negative gradient of the potential energy, the field is called conservative. In such cases, the equation of motion can be written as

$$\text{3D}: \quad m\frac{d^2\mathbf{r}}{dt^2} = -\nabla V(\mathbf{r}); \quad \text{1D}: \quad m\frac{d^2x}{dt^2} = -\frac{dV(x)}{dx} \tag{A.9}$$

There are four basic interactions. The so-called weak and strong interactions are only significant for the components of the atomic nucleus, while gravitation plays an important role only if huge bodies (e.g., stars or planets) are involved. The interaction determining the "everyday life of an engineer" is the Coulomb force between two charges, q and Q:

$$F_C = \frac{1}{4\pi\varepsilon_0}\frac{qQ}{r^2} \tag{A.10}$$

The force acting on q is exerted by the electric field of Q. This field can be characterized by the field strength or by the potential (force or potential energy per unit charge, respectively):

$$|\mathbf{E}| = \frac{1}{4\pi\varepsilon_0}\frac{Q}{r^2}; \quad \Phi = -\frac{1}{4\pi\varepsilon_0}\frac{Q}{r} \tag{A.11}$$

A.4 Examples

A.4.1 Electrons in a Homogenous Electric Field

A constant electric field of strength $\mathbf{E} = E_x\mathbf{e}_x$ gives rise to a dielectric (constant) current because the force of the field is compensated by a friction force, proportional to the velocity:

$$m\ddot{x} = F_x = eE_x - \frac{m}{\tau}\dot{x} = 0 \rightarrow \dot{x} = v_x = \frac{e\tau}{m}E_x \tag{A.12}$$

The friction is determined by the mean scattering time, which is inversely proportional to the temperature: $\tau \sim T^{-1}$.

The current I is the charge transfer in unit time, and the current density j is the current through unit area. In the time Δt, the electrons travel the distance $l = v_x\Delta t$. If the electron density (number per volume) is $n = N/V$, the number of electrons (with charge e each) going through an area A is $N = nlA$. So the current density is

$$j = \frac{I}{A} = \frac{Q}{A\Delta t} = \frac{enlA}{A\Delta t} = \frac{env_x\Delta t}{\Delta t} = env_x = en\frac{e\tau E_x}{m} = \frac{e^2 n\tau}{m}E_x \tag{A.13}$$

This is Ohm's law in differential form $j = \sigma \cdot E_x$, where the conductivity

$$\sigma \equiv \frac{e^2 n\tau}{m}\left(\equiv \frac{1}{\rho}\right) \tag{A.14}$$

is the inverse of the resistivity ρ. Since τ is inverse proportional to the temperature, the resistivity ρ *increases linearly with T*. The resistance is $R = \rho l/A$.

A.4.2 Harmonic Oscillators

Harmonic oscillation arises when a particle, under the effect of a spring force $F = -Dx$, is deflected from its equilibrium position with the amplitude x_0. Introducing the eigenfrequency (or natural frequency) $\omega_0 \equiv \sqrt{D/m}$, the solution of the equation of motion, $m\ddot{x} = -Dx$, is

$$x(t) = x_0 e^{i\omega_0 t} \tag{A.15}$$

The spring force is the negative gradient of the *harmonic potential*, $V = Dx^2/2$, and the total energy

$$E = T + V = \frac{1}{2}m\omega_0^2 \tag{A.16}$$

is constant.

Superposition of harmonic vibrations in a frequency band $[\omega_0 - \Delta\omega, \omega_0 + \Delta\omega]$ results in a nonharmonic vibration restricted in time (see Figure A.2a).

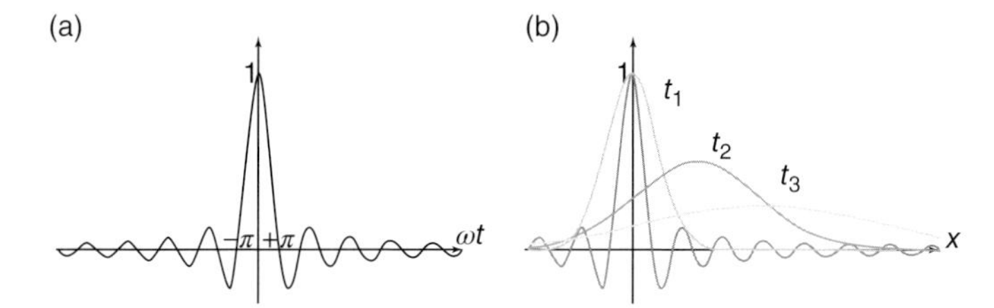

Figure A.2 (a) A finite-time vibration, as a result of the superposition of harmonic vibrations with equal amplitudes in the frequency range $[\omega_0 - \Delta\omega, \omega_0 + \Delta\omega]$. (b) Wave packet (finite wave train), as a result of the superposition of harmonic waves with equal amplitudes in the wavenumber range $[k_0 - \Delta k, k_0 + \Delta k]$. The wave packet delocalizes with time in a dispersive medium.

The product of the vibration time Δt and the frequency half-bandwidth $\Delta\omega$ is constant:

$$\Delta\omega\Delta t \approx 2\pi \tag{A.17}$$

Note that the constant 2π is the difference between the first two nodes of the resulting amplitude, $sin(\Delta\omega t)/(\Delta\omega t)$, which arises if all harmonic vibrations in the superposition have the same amplitude.

A.5 Waves in an Elastic Medium

Newton's axioms in an elastic body, as well as Maxwell's axioms in an electromagnetic field far from charges, lead to a *wave equation*

$$\frac{\partial^2\psi(\mathbf{r},t)}{\partial t^2} = v_f^2 \frac{\partial^2\psi(\mathbf{r},t)}{\partial \mathbf{r}^2} \tag{A.18}$$

The *wave function* ψ determines the state of the medium (e.g., deviation from equilibrium, or value of the field strength) at any given point $\mathbf{r}$ at any given time t. Equation (A.18) is satisfied if ψ contains the two variables in the form: $\mathbf{r} \pm \mathbf{v}_f t$. This means that the points at $\pm\mathbf{r}$ will have the same state after time t as the point at the origin had at the beginning of the motion. In other words, the state propagates with the phase velocity $\mathbf{v}_f$. This phenomenon is called a *wave*.

A local harmonic vibration (with circular frequency $\omega = 2\pi\nu = 2\pi/T$, where ν is the frequency and T the period) in an elastic medium initiates a *harmonic wave* with wavelength $\lambda = v_f(\lambda)T$. In the one-dimensional case,

$$\psi(x,t) = \psi_0 e^{i\left[2\pi\left(\frac{t}{T} - \frac{x}{\lambda}\right) + \alpha\right]} = \psi_0 e^{i(\omega t - kx) + \alpha} \tag{A.19}$$

where $k = 2\pi/\lambda$ is the *wavenumber* and α is a phase constant depending on the initial conditions. In three dimensions, the wavenumber is defined as a vector $\mathbf{k}$, with $2\pi/\lambda$ as its magnitude (norm), and direction parallel to the direction of propagation. In the case of a harmonic wave, the variation in the state of the medium is periodic both in space and in time. The phase velocity may depend on the wavelength (or on the wavenumber) and is determined by the strength of the coupling between volume elements of the medium. The connection between circular frequency and the wavenumber (or between period and wave length) is expressed in the *dispersion relation*

$$\omega(k) = v_f(k) \cdot k \tag{A.20}$$

(or $\lambda = v_f(\lambda)T$). If the phase velocity does not depend on the wavenumber k, the dispersion relation is linear. Nonlinearity of the dispersion relation means that the phase velocity is different for every wavenumber, so waves of different wavenumbers that started together, disperse with time. The *dispersion relation of electromagnetic waves* is

$$\omega(k) = \frac{c}{n(k)} \cdot k \tag{A.21}$$

where $n(k)$ is the index of refraction, which is only constant (= 1) in vacuum.

Coherent harmonic waves ($\omega_1 = \omega_2$), starting in the same direction from the same point in space with a time difference α = constant or at the same time from two different points with distance Δs, interfere with each other and add up to a double amplitude if $\alpha = (2n) \cdot \pi$ or $\Delta s = (2n) \cdot \lambda/2$, respectively ($n$ is an integer). This is known as constructive interference. In contrast, even $(2n + 1)$ multiples lead to zero amplitude, known as destructive interference. A wave reflected from a fixed end of a medium forms a standing wave ψ by interference:

$$\psi = 2\psi_0 \sin k_n x e^{i\omega_n t} \tag{A.22}$$

with nodal points (zero amplitude at any time t) at $x = (2n) \cdot \lambda/4$. *If both ends are fixed, only standing waves with discrete wavenumbers and frequencies are possible:*

$$k_n = \frac{n\pi}{L} \Rightarrow \omega_n = v_f k = v_f \frac{n\pi}{L} \tag{A.23}$$

The superposition of incoherent waves with $\omega_1 \approx \omega_2$ leads to harmonic beats in the resulting amplitude. Superposition of many harmonic waves (infinite in space) with wavenumbers in the band $[k_0 - \Delta k, k_0 + \Delta k]$ leads to a finite, nonharmonic wave train (Figure A.2b), called a *wave packet*. With a suitable choice of the amplitudes $\psi_0(k)$, any arbitrary form of the wave packet can be realized. The wave packet moves with the so-called *group velocity*

$$v_g = \frac{d\omega}{dk} \tag{A.24}$$

From this definition, it follows that in a dispersive medium $v_g \neq v_f$ (cf. Eq. (A.20)).

In a nondispersive medium, the product of the width of the wave packet Δx and the wavenumber half-bandwidth Δk is constant, while in a dispersive medium it increases with time:

$$\Delta k \Delta x \approx 2\pi, \quad \Delta k \Delta x \geq 2\pi, \text{ respectively.} \tag{A.25}$$

that is, the wave packet delocalizes (gets wider and wider, see Figure A.2b). The constant 2π is the difference between the first two nodes of the function $\sin(v_g \cdot t - x)/(v_g \cdot t - x)$, which is the resulting amplitude if the harmonic wave components all have the same initial amplitudes $\psi_0(k)$ = constant.

Real wave sources are never ideally monochromatic ($\Delta\omega \neq 0$), so they actually always emit wave packets. Superposition between the wave packets is only possible if they are emitted within the *coherence time* $\tau_c = 1/\Delta\omega$, or if the difference in the path length to reach a given point is within the *coherence length*: $\lambda_c = v_f \cdot \tau_c$.

A.6 Wave Optics

Maxwell's axioms for the electric field **E** around a charge distribution $\rho(\mathbf{r})$ and for the magnetic induction field **B** around a current density **j** are

$$\nabla\mathbf{E} = \frac{1}{\varepsilon_0}\rho; \quad \nabla \times \mathbf{E} = -\frac{\partial \mathbf{B}}{\partial t}$$

$$\nabla\mathbf{B} = 0; \quad \nabla \times \mathbf{B} = \mu_0\mathbf{j} + \varepsilon_0\mu_0\frac{\partial \mathbf{E}}{\partial t} \tag{A.26}$$

where ε_0 and μ_0 are the electric and magnetic constants, (permittivity and permeability, respectively). Far from charges and currents ($\rho = 0, \mathbf{j} = 0$), Eq. (A.26) reduces to two wave equations (cf. Eq. (A.18)):

$$\frac{\partial^2\mathbf{E}}{\partial t^2} = \frac{1}{\varepsilon_0\mu_0}\Delta\mathbf{E}; \quad \frac{\partial^2\mathbf{B}}{\partial t^2} = \frac{1}{\varepsilon_0\mu_0}\Delta\mathbf{B} \quad \text{with} \quad \frac{1}{\varepsilon_0\mu_0} = c^2 \tag{A.27}$$

The solutions of Eq. (A.27) are the electromagnetic waves, which are termed light if their wavelengths are in the visible region of the spectrum. Light is, therefore, the propagation of vibration in the field strength, and it transports energy. The light energy, transferred by a harmonic plane wave across a unit area during the time Δt,

$$\delta E = I\Delta t = \frac{1}{2}\varepsilon_0 c|\mathbf{E}_0|^2 \tag{A.28}$$

(where I is the intensity), is proportional to the phase velocity of light c and to the absolute square of the amplitude of the field strength. The interference and diffraction phenomena of light can be explained by its wave nature.

A.6.1 Diffraction by a Double Slit

Diffraction by a double slit causes interference bands (dark and light stripes) to appear on a screen set up parallel to the slits. The intensity distribution orthogonal to the optical axis is (see Figure A.3)

$$I \propto \psi_0^2 \frac{\sin^2\beta}{\beta^2}\cos^2\frac{\delta}{2} \tag{A.29}$$

with the width a and distance d of the slits, for the wavelength λ:

$$\beta = \frac{\pi a \sin\alpha}{\lambda}; \quad \delta = \frac{2\pi d \sin\alpha}{\lambda} \tag{A.30}$$

The minima and maxima of the intensity distribution are given in Figure A.3.

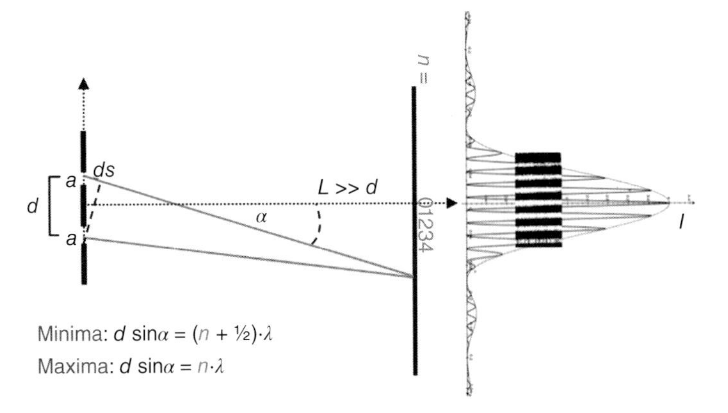

Figure A.3 Diffraction by a double slit.

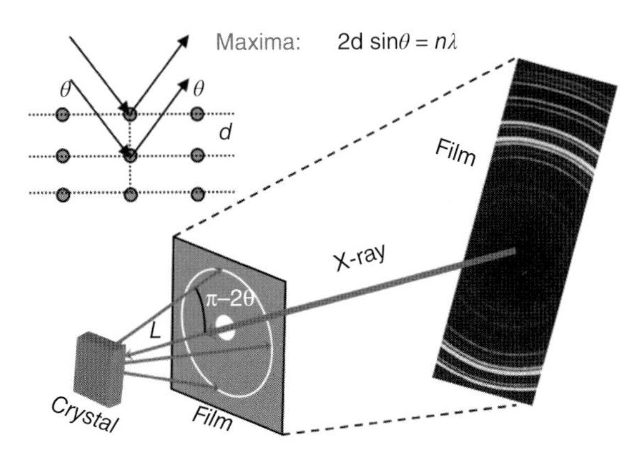

Figure A.4 X-ray diffraction on a polycrystal in the Laue geometry, and the Bragg condition of constructive interference.

A.6.2 X-Ray Diffraction by a Crystal Lattice

X-ray diffraction by a crystal lattice also produces an interference pattern by the superposition of waves reflected from neighboring atomic planes (with a distance d between them). According to the *Bragg equation* shown in Figure A.4, amplification can be expected at the sight angles θ, for which

$$2d \sin \theta = n\lambda \tag{A.31}$$

In the case of a polycrystalline material, where the microcrystals are oriented randomly, the X-ray diffraction in the *Laue geometry* (Figure A.4) gives rise to concentric rings on an X-ray-sensitive film.

A.7 Equilibrium Energy Distribution among Many Particles

The (macro)state of an N-particle system can be characterized by the *extensive thermodynamic parameters*, entropy (S) and volume (V), which determine the total internal energy $U = U(S, V)$. The partial derivatives of this function define the *intensive thermodynamic parameters*, temperature (T) and pressure (p), respectively, as average quantities. The entropy is given by the number w of microstates (different spatial and velocity distributions), which lead to the same macrostate (same U and V at constant N):

$$S = k_B \cdot \ln(w) \tag{A.32}$$

where k_B is the Boltzmann constant. Equilibrium means the macrostate with maximal entropy (logarithm of the maximal probability). It is a combinatorial task to determine the number of microstates that lead to the same macrostate. The maximum belongs to the homogeneous spatial distribution and – depending on the boundary conditions – to one of the following energy distributions.

Distinguishable particles ⇒ **Maxwell–Boltzmann distribution:**

$$f_{\mathrm{MB}}(\varepsilon_i, T) = \frac{e^{-\varepsilon_i/k_B T}}{\sum_i e^{-\varepsilon_i/k_B T}} \tag{A.33}$$

Indistinguishable particles ⇒ **Bose–Einstein distribution:**

$$f_{\mathrm{BE}}(\varepsilon, T) = \frac{1}{e^{\varepsilon_i/k_{\mathrm{B}}T} - 1} \tag{A.34}$$

Indistinguishable particles; two, at most, having the same energy ⇒ **Fermi–Dirac distribution:**

$$f_{\mathrm{FD}}(\varepsilon, T) = \frac{1}{e^{(\varepsilon-\mu)/k_{\mathrm{B}}T} + 1} \tag{A.35}$$

where $f(\varepsilon, T)$ is the probability to find a particle with energy ε under equilibrium conditions at temperature T. In the FD distribution

$$\mu = \frac{\partial(E - TS + pV)}{\partial N} + q\Phi \tag{A.36}$$

which, in the absence of an electric potential Φ in a solid under constant temperature and pressure, can be regarded as the average energy per particle (with charge q).

Knowing the distribution function, the average of any given energy-dependent quantity $X(\varepsilon)$ can be calculated at temperature T as

$$\overline{X}_T = \int X(\varepsilon) f(\varepsilon, T) d\varepsilon \tag{A.37}$$

From the MB Distribution in Eq. (A.33), which pertains to classical particles, it follows that the internal energy of an N-particle system is

$$U = \frac{f}{2} N k_{\mathrm{B}} T \tag{A.38}$$

where $f = 3$ is the number of thermodynamic degrees of freedom (number of independent terms in the total energy expression) per particle. In classical thermodynamics, Eq. (A.38) was generalized for multiatom molecules, that is, it has been assumed that each thermodynamic degree of freedom has

$$\langle \varepsilon \rangle = \frac{1}{2} k_{\mathrm{B}} T \tag{A.39}$$

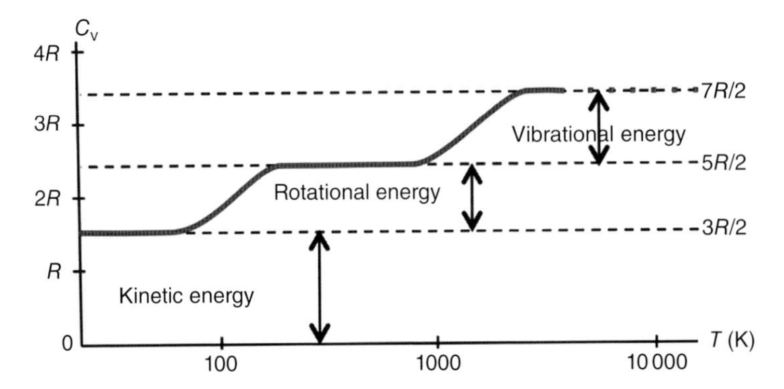

Figure A.5 Molar heat capacity, $C_V \sim dU/dT$, of the H_2 gas as a function of temperature. (N.B.: Two-atomic molecules have three translational, two rotational, and two vibrational degrees of freedom. Heating the gas activates the latter only if a critical energy is reached.) Here, $R = N_A k_B$ is the ideal gas constant, and N_A is Avogadro's number.

energy on average. However, experiments show that this assumption is only valid at rather high temperatures. The various degrees of freedom are activated step by step upon increasing the temperature (see Figure A.5 for the H_2 gas).

A.8 Complementary Variables

To any quantity describing a particle with respect to a frame of reference, for example, x, y, or r, ϕ, or t, that is, to *generalized coordinates* q_i, one can assign a *generalized momentum* p_i through the equation

$$p_i = \frac{\partial(T - V)}{\partial \dot{q}_i} \tag{A.40}$$

where T and V are the kinetic and potential energy, respectively. This way, the momentum $p_x = mv_x$ is complementary to x, $p_y = mv_y$ to y, the angular momentum component $L_z = (\mathbf{r} \times \mathbf{p})_z$ to ϕ, and the negative of the total energy $-E$ to t. The product of the units of complementary pairs is always [J s]. The generalized coordinates and momenta fulfill the equations

$$\dot{q}_i = \frac{\partial H}{\partial p_i}; \quad \dot{p}_i = -\frac{\partial H}{\partial q_i} \tag{A.41}$$

where

$$H = T + V \tag{A.42}$$

is the *Hamiltonian function* of a conservative system. Equation (A.41) represents an alternative to the Newtonian equation of motion.

A.9 Special Relativity Theory

The contradiction between the facts that the speed of a body is dependent on the chosen frame of reference but that of light is not, can be resolved by the *Lorentz transformation* between frames of reference:

$$\begin{bmatrix} ct \\ x \\ y \end{bmatrix} = \begin{pmatrix} \gamma & \beta\gamma & 0 \\ \beta\gamma & \gamma & 0 \\ 0 & 0 & 0 \end{pmatrix} \begin{bmatrix} ct' \\ x' \\ y' \end{bmatrix} \tag{A.43}$$

where v is the uniform relative speed between the systems, c is the speed of light, $\beta \equiv v/c$, and $\gamma = 1/(\sqrt{1-(v^2/c^2)})$. This transformation is based on the realization by *Einstein* that not only the coordinates but also the time is different in two frames of reference. From Eq. (A.43) it follows that the length of an object and the time span of an event in the frame of reference with origin O' are both different from those in the system with origin O. If the O-system is regarded as being at rest,

$$\Delta x = \Delta x' \sqrt{1 - v^2/c^2}; \quad \Delta t = \frac{\Delta t'}{\sqrt{1 - v^2/c^2}} \tag{A.44}$$

The invariant quantity is the "distance" in the *space–time continuum:*

$$x^2 + y^2 + z^2 + (ict)^2 = x'^2 + y'^2 + z'^2 + (ict')^2 \tag{A.45}$$

Laws of physics must be invariant to the Lorentz transformation. To achieve that in the case of the principle of momentum (see Eq. (A.6)), the inertia m of a point mass, in the momentum $p = mv$, must depend on the speed

$$m = \frac{m_0}{\sqrt{1 - v^2/c^2}} \tag{A.46}$$

This means an infinite mass m if $v = c$. In other words, a body with finite m_0 mass at rest (*rest mass*) cannot achieve the speed of light, for that would require infinite force. It can also be shown that bodies with rest mass m_0 incorporate energy

$$E_0 = m_0 c^2 \tag{A.47}$$

even if they do not move with respect to the chosen frame of reference. The total energy of a body is

$$E = \sqrt{(pc)^2 + (m_0 c^2)^2} \tag{A.48}$$

plus the potential energy. Equation (A.47) expresses the equivalence of mass and energy, which can indeed be transformed into each other (see, e.g., nuclear fission or fusion).

B

Important Mathematical Formulas

B.1 Numbers

The *complex plane* is defined by the orthogonal axes of the real and imaginary numbers. The points in this plane are complex numbers, $z = x + iy$, where $i = \sqrt[+]{-1}$ and $1/i = -i$. The addition of complex numbers means adding up the real and imaginary parts separately: $z_1 + z_2 = (x_1 + x_2) + i(y_1 + y_2)$. Multiplication is done in a simple, distributive manner: $z_1 - z_2 = (x_1x_2 - y_1y_2) + i(x_1y_2 - y_1x_2)$. The number $z^* = x - iy$ is called the *complex conjugate of* z. The product $z^*z = x^2 + y^2$ is the absolute square, and the square root of that is the norm $|z|$ of the complex number. Division by a complex number is equivalent to multiplying with the complex conjugate and dividing by the absolute square. The complex conjugate of an exponential function means the complex conjugation of the argument. A number on the complex plane can also be given in a trigonometric form, $z = |z|(\cos\phi + i\sin\phi)$, where $\tan\phi = y/x$. According to the Euler formula, this is equivalent to $z = |z|\cdot\exp\{i\phi\}$, and it follows that

$$\cos\phi = \frac{e^{i\cdot\phi} + e^{-i\cdot\phi}}{2}, \quad \sin\phi = \frac{e^{i\cdot\phi} - e^{-i\cdot\phi}}{2i} \tag{B.1}$$

Number series can be constructed by simple rules, and the sum of the series up to a given element can be given with the following formulas:

$$\text{Arithmetic series,}\ a_n = a_1 + (n-1)d \quad \sum_{i=1}^{n} a_i = \frac{n(a_1 + a_n)}{2} \tag{B.2}$$

$$\text{Geometrical series,}\ a_n = a_1 q^{(n-1)} \quad \sum_{i=1}^{n} a_i = \frac{a_1(1-q^n)}{1-q} \tag{B.3}$$

$$\text{Factorial,}\ n! = 1\cdot 2\cdot 3\cdot\dots\cdot n, \quad \sum_{n=0}^{\infty} \frac{1}{n!} = e \tag{B.4}$$

(Here, d and q are constants.)

For any arbitrary set of numbers, a weighted average can be defined. If a given number $k_i(i = 1,2,\dots)$ occurs N_i times in the set, the weighted average or *mean value* is

$$\langle k\rangle = \frac{1}{N}\sum_i N_i k_i; \quad N = \sum_i N_i \tag{B.5}$$

Essential Quantum Mechanics for Electrical Engineers, First Edition. Peter Deák.

The square root of the average quadratic deviation from the mean value is the standard deviation:

$$\Delta k = \left[\frac{1}{N}\sum_{i=1}^{N}(k_i - \langle k \rangle)^2\right]^{1/2} = \sqrt{\langle k^2 \rangle - \langle k \rangle^2} \tag{B.6}$$

B.2 Calculus

Analytic functions are mathematical instructions for assigning numbers in one mathematical set to the numbers in another by a series of elementary operations (addition, multiplication, sign change, inversion). If numbers (or vectors) are assigned to vectors, we talk about a scalar (or vector) field.

Calculus is the part of mathematics dealing with derivatives and integrals. *Derivatives* describe the changes of a function. The first derivative gives the slope, and the second gives the curvature as a function of the variables. In the case of fields, partial derivatives have to be calculated, that is, derivatives with respect to each coordinate, while keeping the others constant. The total derivative is then

$$df(\mathbf{r}) \equiv \frac{\partial f(\mathbf{r})}{\partial x}dx + \frac{\partial f(\mathbf{r})}{\partial y}dy \tag{B.7}$$

The slope of a scalar field is given by the gradient vector

$$\frac{df(\mathbf{r})}{d\mathbf{r}} \equiv \frac{\partial f(\mathbf{r})}{\partial x}\mathbf{e}_x + \frac{\partial f(\mathbf{r})}{\partial y}\mathbf{e}_y = \left[\frac{\partial}{\partial x}\mathbf{e}_x + \frac{\partial}{\partial y}\mathbf{e}_y\right]f(\mathbf{r}) \equiv \nabla f(\mathbf{r}) \tag{B.8}$$

where ∇ is called the *nabla operator*. The second derivative is a scalar defined by the so-called Laplace operator Δ:

$$\frac{d^2 f(\mathbf{r})}{d\mathbf{r}^2} \equiv \nabla[\nabla f(\mathbf{r})] = \frac{\partial^2 f(\mathbf{r})}{\partial x^2} + \frac{\partial^2 f(\mathbf{r})}{\partial y^2} \equiv \Delta f(\mathbf{r}) \tag{B.9}$$

The derivatives of the elementary functions are

$$\frac{d}{dx}x^n = nx^{n-1}, \quad \frac{d}{dx}e^x = e^x, \quad \frac{d}{dx}\ln(x) = \frac{1}{x}, \quad \frac{d}{dx}\sin x = \cos x, \quad \frac{d}{dx}\cos x = -\sin x \tag{B.10}$$

Rules of derivation:

$$\text{Sum rule}: \quad \frac{d}{dx}[f(x) + g(x)] = \frac{df(x)}{dx} + \frac{dg(x)}{dx} \tag{B.11}$$

$$\text{Product rule}: \quad \frac{d}{dx}[f(x)g(x)] = \frac{df}{dx}g + f\frac{dg}{dx} \tag{B.12}$$

$$\text{Quotient rule}: \quad \frac{d}{dx}\left[\frac{f(x)}{g(x)}\right] = \frac{f'g - fg'}{g^2} \tag{B.13}$$

$$\text{Chain rule}: \quad \frac{d}{dx}f[y(x)] = \frac{df}{dy}\frac{dy}{dx} \tag{B.14}$$

The *integral* of a function is the weighted sum of the function values over a given range of the variable, and the weighting factors are the intervals of infinitesimal size into which the range is divided. In a field, one can integrate along a line, over an area or volume. The value of the integral is obtained (in 1D between $x = a$ and b) with the help of the primitive function (or antiderivative) F:

$$\int_a^b f(x)dx = [F(b) - F(a)] \equiv [F(x)]_a^b \tag{B.15}$$

The primitive function of $f(x)$ is the function $F(x)$ for which $dF(x)/dx = f(x)$. (This explains the name "antiderivative.") If $F(x)$ is one primitive function of $f(x)$, then all other functions, obtained by adding an integer to $F(x)$, are also primitive functions. All such primitives together are called the *indefinite integral* (which is the inverse operation to derivation):

$$\int f(x)dx = F(x) + C \tag{B.16}$$

An arbitrary function $f(x)$ can be *expanded on the basis of a complete set of known functions* $\{f_n(x)\}$, that is, it can be expressed as a linear combination of the elements in the set. (The set $\{f_n(x)\}$, is complete if none of its elements can be expressed as linear combination of the others, and if there is no function outside the set that cannot be expressed as a linear combination.) The most often applied expansions use exponential or power functions as basis:

$$\text{Fourier-expansion}: \quad f(x) = \sum_{k=-\infty}^{\infty} f_k e^{ikx} \tag{B.17}$$

$$\text{Power expansion}: \quad f(x) = \sum_{n=0}^{\infty} a_n x^n \tag{B.18}$$

For example, the power expansion of the exponential function is

$$e^x = 1 + x + \frac{x^2}{2!} + \frac{x^3}{3!} + \cdots \tag{B.19}$$

The Taylor series is a power expansion with coefficients determined by the derivatives.

$$f(x) = f(0) + \left(\frac{df}{dx}\right)_0 x + \left(\frac{d^2f}{dx^2}\right)_0 \frac{x^2}{2!} + \left(\frac{d^3f}{dx^3}\right)_0 \frac{x^3}{3!} + \cdots \tag{B.20}$$

The Taylor series is very convenient for approximating a complicated known function $f(x)$ in the vicinity of $x = 0$ with the first few terms of the expansion.

B.3 Operators

Operators are instructions for assigning one function to another by means of mathematical operations. For example,

Multiplication by a variable	$x \cdot f(x, y)$	→ assigns, for example, x^3 to x^2
Absolute square formation	$\|f(x, y)\|$	→ assigns, for example, x^2 to x^2
Square root formation	$\sqrt{f(x, y)}$	→ assigns, for example, $\pm x$ to x^2
Partial derivation	$\frac{\partial}{\partial x} f(x, y)$	→ assigns, for example, $2x$ to x^2
Integration	$\int f(x, y) d\mathbf{r}$	→ assigns, for example, $x^3/3$ to x^2

Operators are often abbreviated by a letter with a hat, for example,

$$\frac{d}{dx} f(x) \equiv \hat{D}f; \quad \int f(x)dx \equiv \hat{I}f; \quad x \cdot f(x) \equiv \hat{x}f$$

In general, $\hat{O}$ (pronounce: "O-operator") means some given operation on a function.

Eigenfunctions of $\hat{O}$ are those functions for which the effect of the operation corresponds to a multiplication with a constant. For example, the exponential functions are eigenfunctions of the differential operator:

$$\frac{d}{dx} e^{kx} = ke^{kx} \tag{B.21}$$

The constant, with which the eigenfunction is multiplied, is called the *eigenvalue*. Different eigenfunctions may have the same eigenvalue. For example, for the operator $\hat{D}^2 \equiv d^2/dx^2$ both $exp(\sqrt{k}x)$ and $exp(-\sqrt{k}x)$ are so-called *degenerate eigenfunctions* with the same eigenvalue k:

$$\frac{d^2}{dx^2} e^{\pm\sqrt{k}x} = ke^{\pm\sqrt{k}x}$$

In order to find all possible ($i = 1, 2, 3, \ldots$) eigenfunctions and their eigenvalues of a given operator $\hat{O}$, the *eigenvalue equation*,

$$\hat{O}f_i = k_i f_i \tag{B.22}$$

has to be solved. In the case of $\hat{D}^n$, this is a differential equation.

B.4 Differential Equations

Differential equations (DEs) are equations for the determination of an unknown function $f(x)$. The DE contains derivatives of that function and other known functions. The *order* of the DE is the order of the highest derivative of f that occurs in the equation. In the case of 1D functions, the DE is called *ordinary*; if partial derivatives occur, it is a *partial* DE. The standard form of an ordinary DE of second order (ODE^2) is

$$P(x)\frac{d^2 f(x)}{dx^2} + Q(x)\frac{df(x)}{dx} + R(x)f(x) + S(x) = 0 \Rightarrow f(x) = ?? \tag{B.23}$$

where $P(x)$, $Q(x)$, $R(x)$, and $S(x)$ are known functions and $f(x)$ is the solution. Simple ODE^2 can be solved heuristically. In complicated cases, the solution can be found with the help of an expansion over a known, complete set of functions, for example,

$$f(x) = e^{-x} \sum_m c_m x^m$$

The ODE[2] can then be converted into a first-order system of algebraic equations, from which the coefficients can be determined.

It is important to note that if a complex-valued function $f(z)$ is a solution to a DE, then the complex conjugate $f^*(z)$ is also a solution.

The general solution of a DE can be written as the linear combination of two special solutions. The coefficients can be determined with the help of the boundary conditions.

B.5 Vectors and Matrices

Vectors and matrices are, in a generalized sense, one- and multidimensional data fields, respectively, where each data position is assigned to one specific quantity. The data fields can be added up by adding the data in the corresponding positions. In the case of 3D geometric space, the positions are assigned to the coordinates along the Cartesian axes, and

$$\mathbf{a} + \mathbf{b} \equiv \sum_{i=1}^{3} (a_i + b_i)\mathbf{e}_i \tag{B.24}$$

where $\mathbf{e}_i$ are the unit vectors along the axes. Between such vectors, one can define a so-called scalar product (or dot product):

$$\mathbf{a} \cdot \mathbf{b} = \sum_{i=1}^{3} a_i b_i; \quad |\mathbf{a} \cdot \mathbf{b}| = |\mathbf{a}||\mathbf{b}| cos\, \gamma \tag{B.25}$$

and a vector product (or cross product):

$$\mathbf{a} \times \mathbf{b} = \det \begin{pmatrix} \mathbf{e}_1 & \mathbf{e}_2 & \mathbf{e}_3 \\ a_1 & a_2 & a_3 \\ b_1 & b_2 & b_3 \end{pmatrix}; \quad |\mathbf{a} \times \mathbf{b}| = |\mathbf{a}||\mathbf{b}| \sin \gamma \tag{B.26}$$

where γ is the angle between the vectors. The determinant of a 3×3 matrix is

$$\det \begin{pmatrix} A_{11} & A_{12} & A_{13} \\ A_{21} & A_{22} & A_{23} \\ A_{31} & A_{32} & A_{33} \end{pmatrix} = A_{11}(A_{22}A_{33} - A_{32}A_{23}) - A_{12}(A_{21}A_{33} - A_{31}A_{23}) + A_{13}(A_{21}A_{32} - A_{31}A_{22}) \tag{B.27}$$

C

List of Abbreviations

In this book, vectors are denoted in bold and scalars in italics: for example, the magnitude of a vector is $|\mathbf{A}| = A$. Operators are given with a hat over the letter. Unless noted otherwise, the letters can have one of the following meanings (corresponding to the text environment):

	Vectors		Scalars (magnitude)		Vectors		Scalars (magnitude)
a		*a*		**A**		*A*	
	Acceleration		Acceleration		Area vector normal to surface		Area
	General vector		General real number		Acceleration		General real number
b		*b*		**B**		*B*	
	General vector		General real number		Magnetic induction		Magnetic induction
			Penetration depth				
c		*c*		**C**		*C*	
	General vector		General real number				Capacity
			Speed of light				Conduction band edge
			Concentration				Molar heat
			Specific heat				Constant
d		*d*		**D**		*D*	
			Diameter or distance		Electric displacement		Electric displacement
			Total derivative				Spring constant
							Dimension
e		*e*		**E**		*E*	
	Unit vector		2.71828182845905		Electric field strength		Electric field strength
			Energy of a single particle				Total energy

Vectors		Scalars (magnitude)		Vectors		Scalars (magnitude)	
f		f	General function Number of the degrees of freedom Occupation probability	**F**	Force	F	Force
g		g	Free fall acceleration General function	**G**		G	
h		h	Height Planck's constant ($\hbar = h/2\pi$)	**H**		H	Hamilton function
i		i	Imaginary unit Index for integers	**I**		I	Current Intensity
j	Current density	j	Current density Index for integers	**J**		J	
k	Wavenumber	k k_B	Wavenumber Index for integers Constant, eigenvalue Boltzmann constant	**K**		K	Constant
l		l	Index for integers Angular momentum quantum number Length	**L**	Angular momentum	L	Angular momentum Length
m		m	Index for integers Mass Magnetic quantum number	**M**	Torque	M	Mass Torque
n	$[n_1, n_2, n_3]$	n	Index for integers Number density Number of moles Index of refraction	**N**		N N_A	Number of particles Avogadro's number
o		o		**O** $\hat{O}$	General operator	O	

	Vectors		Scalars (magnitude)		Vectors		Scalars (magnitude)
p	(Linear) momentum Generalized momentum	*p*	(Linear) momentum Generalized momentum Pressure	**P**	Total (linear) momentum Polarization	*P*	Total (linear) momentum Polarization Power
q	Generalized coordinate	*q*	Generalized coordinate Charge, electron charge	**Q**		*Q*	Charge Heat
r	Position vector	*r*	Radius	**R**	Position vector	*R*	Radius Resistance Universal gas constant
s	Displacement	*s*	Length of path	**S**		*S*	Entropy Overlap
t		*t*	Time	**T**		*T*	Period Temperature Kinetic energy Transmission
u		*u*	Energy density	**U**		*U*	Bias Internal energy
v	Speed	v	Speed	**V**	Speed	*V*	Speed Volume Potential energy Valence band edge
w		*w*	TD probability	**W**		*W*	Work Probability
x	Position vector $[x_1, x_2, x_3]$	*x*	Variable, coordinate	**X**		*X*	Coordinate
y		*y*	Variable, coordinate	**Y**		*Y*	Coordinate
z		*z*	Variable, coordinate Complex number	**Z**		*Z*	Coordinate State sum (partition function)

Greek and mathematic symbols:

α	Angle Phase factor Constant			∇	Nabla operator $(=d/d\mathbf{r})$
β	Angle Angular acceleration v/c Constant			∂	Partial derivative
γ	Angle Damping constant $\left(1-\frac{v^2}{c^2}\right)^{-1}$	Γ		$\times$	Vector product
δ	Difference	Δ	Difference Laplace operator Uncertainty, standard deviation	$\langle\,\vert\,\rangle$	Scalar product of two functions
ε	Dielectric constant Energy of a single particle			$< >$	Mean value, expectation value
η	Efficiency				
θ	Angle, polar angle	Θ	Momentum of inertia	f'	Space derivative of f (df/dx)
ϑ	Angle, polar angle			$\dot{x}$	Time derivative of x (dx/dt)
λ	Wavelength Mean free path	Λ	Mean free path		
μ	Chemical potential of electrons Magnetic permeability Mobility				
ν	Frequency				
π	3.14159265358979	Π	Multiplication		
ρ	Resistivity Mass density Charge density				

σ	Conductivity	Σ	Summation
τ	Relaxation time Mean scattering time		
φ	Eigenfunction Position-dependent part of the wavefunction		
ϕ	Angle, azimuth angle Phase factor	Φ	Potential Position-dependent part of a many-body wavefunction Flux, radiant flux
ψ	Wavefunction	Ψ	Many-body wavefunction
ω	Angular velocity or circular frequency	Ω	Spatial angle Angular velocity

Solutions

Chapter 1

Exercise 1.1 *Which kinematic and dynamic quantities can be used to characterize the motion of a point mass?*
Kinematic: position, velocity, acceleration (as functions of time).
Dynamic: linear and angular momentum, energy (as functions of time and/or position).

Exercise 1.2 *How are canonically conjugate coordinates (q) and dynamic quantities (p) related to each other?*
Through the Lagrange function, $L = T - V: p = \partial L/\partial \dot{q}$.

Exercise 1.3 *What is the basic assumption behind Ohm's law?*
Electrons are scattered by the nuclei, so the field can only accelerate them during a mean scattering time, up to a constant velocity.

Exercise 1.4 *Demonstrate that a harmonic wave satisfies the wave equation!*
Inserting the second derivatives of the wave function, with respect to position and time, into the wave equation, leads to the dispersion relation.

Exercise 1.5 *What is the dispersion relation of light (i.e., of the EM field) in vacuum? Is the phase and the group velocity equal in this case?*
Since the dispersion relation, $\omega = ck$, is linear, yes!

Exercise 1.6 *What is stated in the equipartition theorem of Boltzmann?*
At temperature T on temporal average, each thermodynamic degree of freedom stores $k_B T/2$ energy.

Chapter 2

Exercise 2.1 *By what factor should the electric power be increased to raise the temperature of the wire in a light bulb from 2000 to 2500 °C? Let us assume that the wire is heated by 100% of the electrical energy, and can be regarded as ideal radiator, following the law of Stefan and Boltzmann.*
By a factor of 2.22.

Essential Quantum Mechanics for Electrical Engineers, First Edition. Peter Deák.

Exercise 2.2 *The surface temperature of the star Aldebaran is 4100 °C. Estimate the wavelength λ_m at which the most energy is radiated. What is the color of Aldebaran?*
$\lambda_m = 662.2$ nm (red).

Exercise 2.3 *Explain why Planck's law is in contradiction with the concepts of classical physics.*
The energy of light should be proportional to the phase velocity and to the square of the field-strength amplitude but not to the frequency. The energy of the vibrating metal atoms ($E = m\omega_0{}^2 x_0{}^2/2$) should also be a continuous quantity.

Chapter 3

Exercise 3.1 *Consider the interaction of light (traveling in vacuum) and a material with two well-defined energy states (two-level system). The photon energy is 3 eV.*

(a) *What is the light frequency? Calculate the wavelength from the dispersion relation. What is the momentum of the photons?*
$\nu = 0.72 \times 10^{15}$ Hz; $\lambda = 413$ nm; $p = 1.6 \times 10^{-27}$ kg m s^{-1}.

(b) *If the absorption constant is $B_{ij} = 10^9$ [m·kg^{-1}], what are the constants of the spontaneous and of the stimulated emission, A_{ji} and B_{ji}, respectively? (Hint: compare Eqs (3.7) and (3.4)!)*
$A_{ji} = 2.31 \times 10^{-4}$ Hz; $B_{ij} = 10^9$ m·kg^{-1}.

(c) *Demonstrate that population inversion is not possible in the thermal equilibrium of a two-level system. (Hint: Use the Maxwell–Boltzmann distribution of Section A.7!)*

$N_i/N_j > 1$ since $E_j > E_i$.

Exercise 3.2 *An X-ray with the wavelength of 5×10^{-12} m is scattered by electrons. The recoiling electron has the energy of 70 keV. Under what angle will the photon be reflected and what will be its wavelength?*
Due to the conservation of energy: $\lambda = 6.97 \times 10^{-12}$ m, and from the Compton formula: $\theta = 79.1°$.

Exercise 3.3 *What is meant by the wave–particle duality of light? What experiments have made this assumption necessary?*
Light can behave as a wave but also as a point-mass-like particle, because it shows interference (e.g., in the double-slit experiment of Young) but, if colliding with an electron, it transfers a frequency-dependent energy and a wave number-dependent momentum (photoelectric and photovoltaic effects, Compton effect).

Chapter 4

Exercise 4.1 *According to Bohr's model of the hydrogen atom, what wavelength should a photon have to*

(a) *excite an electron from the ground state into the third allowed energy level?*
$\lambda = 102.5\,\text{nm}$

(b) *to ionize the atom when the electron is already in this excited state?*
$\lambda = 820.0\,\text{nm}$.

Exercise 4.2 *Apply Bohr's formulas to the He^+ ion.*

(a) *Write down the equations for the radius and energy of the orbitals, and calculate the second smallest radius.*
$r = 0.1058\,\text{nm}$.

(b) *What is the relation between the frequencies of the hydrogen spectrum and of the spectrum of He^+?*
$\nu_{\text{H}}/\nu_{\text{He}} = 0.25$.

Exercise 4.3 *Consider the helium atom in the framework of the Bohr model.*

(a) *What would be the energy of a He atom in the ground state (two electrons in orbital $n = 1$), if the Coulomb repulsion between the electrons could be neglected?*
$E_0 = -108.85\,\text{eV}$.

(b) *Estimate the effect of the Coulomb repulsion between the electrons, $V_C = e^2/(4\pi\varepsilon_0 r_{12})$, assuming that their distance r_{12} is the longest possible on the $n = 1$ orbital. What would be the total energy of the ground state with this estimate? (N.B.: Comparison with the experimental value of $-78.9\,\text{eV}$ shows the error of the Bohr model.)*
$V_C = 27.21\,\text{eV}$; $E = -81.64\,\text{eV}$.

Exercise 4.4 *The quantization of the angular momentum can be derived from the assumption that the electron wave must be stationary on a circular orbit. Use this assumption and the Bohr radii of hydrogen to calculate the momentum and frequency of the electron on the first two orbitals ($n = 1$ or 2).*
$p_1 = 2 \times 10^{-24}\,\text{kg m s}^{-1}$; $\nu_1 = 3.3 \times 10^{15}\,\text{Hz}$
$p_2 = 1 \times 10^{-24-}\text{kg m s}^{-1}$; $\nu_2 = 0.8 \times 10^{15}\,\text{Hz}$.

Exercise 4.5 An electron beam, accelerated by the voltage $U = 63.64\,\text{V}$, is diffracted by a copper crystal. A diffraction maximum ($n = 1$) is observed under the angle of 25°. What is the distance between the reflecting atomic planes?
$d = 1.82\,\text{Å}$.

Exercise 4.6 *What does the wave–particle duality principle say and what experiments have made its formulation necessary?*
The electron can behave both as a point-mass-like particle and as a wave. The electron with its impartible charge (Millikan experiment) is the embodiment of a point mass, still electron diffraction on crystals can be observed (Davisson–Germer experiment).

Chapter 5

Exercise 5.1 *Describe the two main concepts of classical physics for the different forms of matter!*
Point mass (completely localized)
EM wave (delocalized).

Exercise 5.2 *How can the double-slit experiment with single electrons be explained, based on the Born–Jordan interpretation of the concept of a subatomic particle?*
Position measurement localizes the particle but this is a probability event, with the probability distribution provided by the absolute square of the electron wave function (i.e., the intensity).

Exercise 5.3 *What determines the state of a particle in quantum mechanics?*
The wave function contains all information about the state of the particle, and it can be determined from the Schrödinger equation if the potential energy function is known.

Exercise 5.4 *Considering the uncertainty principle of Heisenberg, what quantum mechanical states come closest to the classical concepts of a point mass and a wave?*
Δx Small, Δp large, and Δx large, Δp small, respectively.

Chapter 6

Exercise 6.1 *The normalized wave function of an electron, ψ, can be written as the linear combination of two normalized eigenfunctions, φ_1 and φ_2, of the operator $\hat{O}$ assigned to the measurement: $\psi = a(5\varphi_1 + 12\varphi_2)$. The eigenvalues corresponding to these eigenfunctions are k_1 and k_2, respectively.*

(a) *Determine the real number a!*
$a = 1/13$.

(b) *What are the possible outcomes of the measurement?*
k_1 and k_2.

(c) *What are the probabilities of these results?*
25/169 and 144/169.

(d) *What is the expectation value of this measurement?*
$<O> = (25k_1 + 144k_2)/169$.

Exercise 6.2 *The repeated measurement of the energy of a classical particle as a function of time (under the same conditions) resulted in the values shown in the following table:*

Time	Measurement 1	Measurement 2	Measurement 3
t_1	1.0	1.1	0.9
t_2	2.0	1.9	1.9
t_3	3.0	3.2	3.0

Plot the mean value of the measurements as a function of time and provide error bars corresponding to the standard deviation, $\pm\sigma$. Round up results to two decimal digits and use Eqs (B.5) and (B.6))!
t_1: $\bar{E} = 1.00$; $\Delta E = \pm 0.04$
t_2: $\bar{E} = 1.98$; $\Delta E = \pm 0.02$
t_3: $\bar{E} = 3.05$; $\Delta E = \pm 0.04$.

Exercise 6.3 *From the viewpoint of the observable O, when is the particle in an eigenstate and when in a mixed state? What can the QM say about the result of measuring O in these two cases? What happens if a measurement is attempted on a mixed state?*
If ψ is an eigenfunction of the operator $\hat{O}$, then the particle is in an eigenstate of O, otherwise it is in a mixed state. In the former case, the outcome of the measurement can be accurately predicted: it will be the eigenvalue belonging to the eigenfunction. In the latter case, only mean value and uncertainty (standard deviation) of many measurements can be predicted, and any single measurement yields a random eigenvalue. After the measurement the particle will be in the eigenstate belonging to the measured eigenvalue.

Exercise 6.4 *What are the consequences of the von Neumann theorem for the unambiguous characterization of a state with two observables A and B?*
If the operators of the observables are commutable, then there is at least one state that is simultaneously an eigenstate of both operators and can, therefore, be characterized by both observables accurately. If they are not, at most only one of them can be measured accurately.

Chapter 7

Exercise 7.1 *Under what conditions can two observables be accurately measured in the same state simultaneously?*
When the corresponding operators are commutable.

Exercise 7.2 *Are the operators of the various coordinates, x, y, and z commutative? Check it on a function $\varphi(x, y, z)$!*
Yes. Multiplication with real numbers is commutative.

Exercise 7.3 *Demonstrate that the operators of p_x and y are commutative!*
Yes, because $\frac{\partial y}{\partial x} = 0$.

Exercise 7.4 *Prove that the operator for the absolute square of the angular momentum, and the operator of its z-component are commutative!*
$[\hat{L}^2, \hat{L}_z] = [(\hat{L}_x^2 + \hat{L}_y^2 + \hat{L}_z^2), \hat{L}_z]$ and, of course, $[\hat{L}_z^2, \hat{L}_z] = 0$. Using $[\hat{L}_i, \hat{L}_j] = i\hbar\hat{L}_k$ (see Eq. (7.10)) and by adding an extra term of zero value:

$$[\hat{L}_x^2, \hat{L}_z] = \hat{L}_x\hat{L}_x\hat{L}_z + (-\hat{L}_x\hat{L}_z\hat{L}_x + \hat{L}_x\hat{L}_z\hat{L}_x) - \hat{L}_z\hat{L}_x\hat{L}_x$$
$$= \hat{L}_x[\hat{L}_x, \hat{L}_z] + [\hat{L}_x, \hat{L}_z]\hat{L}_x = -i\hbar(\hat{L}_x\hat{L}_y + \hat{L}_y\hat{L}_x)$$
$$[\hat{L}_y^2, \hat{L}_z] = \hat{L}_y\hat{L}_y\hat{L}_z + (-\hat{L}_y\hat{L}_z\hat{L}_y + \hat{L}_y\hat{L}_z\hat{L}_y) - \hat{L}_z\hat{L}_y\hat{L}_y$$
$$= \hat{L}_y[\hat{L}_y, \hat{L}_z] + [\hat{L}_y, \hat{L}_z]\hat{L}_y = +i\hbar\hat{L}_y(\hat{L}_y\hat{L}_x + \hat{L}_x\hat{L}_y)$$

Exercise 7.5 *What is the operator for the x-component of the angular momentum?*
$\hat{L}_x = \hat{y}\hat{p}_z - \hat{p}_y\hat{z}$

Exercise 7.6 *What consequence does the uncertainty principle have for the Bohr orbitals?*
If the electron has a given angular momentum, for example, in the z-direction, its position along the circular Bohr orbital cannot be determined.

Exercise 7.7 *What is the current density delivered by a single electron, provided that its wave function is $\psi(x,t) = (1/\sqrt{L})e^{i(kx-\omega t)}$?*
With the elementary charge $q = -e$ and the electron mass m_e:

$$\langle j_x \rangle = q\langle \mathrm{v}_x \rangle = \frac{q}{me}\langle p_x \rangle = \frac{q\hbar}{im_e}\int_0^L \frac{1}{\sqrt{L}}e^{-i(kx-\omega t)}\frac{\partial}{\partial x}\left[\frac{1}{\sqrt{L}}e^{-i(kx-\omega t)}\right]dx = \frac{q\hbar}{m_e}k$$

Chapter 8

Exercise 8.1 *A particle moving in the potential $V(x) = (1/2)Dx^2 = (1/2)m\omega_0^2x^2$ is called a quantum mechanical oscillator because $x = x_0 sin(\omega_0 t)$ still holds (x_0 is the amplitude of the vibration). Derive a relation between the uncertainty of position, Δx, and the uncertainty of momentum, Δp_x, for this case, using Eq. (7.21) and the trivial trigonometric relations $<sin(\omega_0 t)> = <cos(\omega_0 t)>$ and $<sin^2(\omega_0 t)> = <cos^2(\omega_0 t)>$!*
$\Delta p = m\omega_0\Delta x$

Exercise 8.2 *Using the previous result, determine the dependence of both uncertainties on the properties of the oscillator (m and ω_0), using Eq. (7.3)!*
$\Delta x = \sqrt{\frac{\hbar^2}{2m\omega_0}}$; $\Delta p = \sqrt{\frac{\hbar^2 m\omega_0}{2}}$

Exercise 8.3 *In contrast to the classical harmonic oscillator, the energy of such bound states is quantized. Based on the uncertainties derived in the previous problem, estimate the lowest allowed energy! Why must this be larger than zero?*

$E_0 \approx \Delta E \approx \Delta T + \Delta V = \frac{\hbar\omega_0}{2}$ (In case of $E_0 = 0$, both x and p would have a definite value.)

Exercise 8.4 *Atoms of a solid can be considered to be harmonic oscillators. Using Planck's hypothesis for the energy absorption of a metal wall, estimate the quantum mechanically allowed energies of the atomic vibrations, taking also into account the minimal energy obtained above! If you did everything right, your result will agree with the energy eigenstates obtained by solving the time-independent Schrödinger equation of a quantum mechanical oscillator exactly.*

$$E_n = \left(n + \frac{1}{2}\right)\hbar\omega_0$$

Exercise 8.5 *The solution of the time-independent Schrödinger equation for the harmonic oscillator gives the following ground state:*

$$\varphi_0 = \frac{1}{(\pi x_0^2)^{1/4}} e^{-\frac{1}{2}\left(\frac{x}{x_0}\right)^2}$$

Calculate the probability density of the particle at $x = 2x_0$ and interpret the result from the viewpoint of classical mechanics!

The probability to find the particle within $\pm dx$ around $2x_0$ is $W(2x_0) = |\varphi_0(2x_0)|^2\, dx = 0.076(dx/x_0)$, that is, not zero. In classical mechanics, the maximum amplitude is x_0.

Chapter 9

Exercise 9.1 *Two energy eigenstates in an infinite quantum well have two and four nodes, respectively. If the width of the well is $L = 10$ nm, what is the difference in the emitted photon energy if the electron returns to the ground state from one or the other state?*

The number of nodes determines the energy quantum number. Therefore, $\Delta E = 0.06$ eV.

Exercise 9.2 *The ground state of an electron in an LED can be given as $E_n^e = E_C + (h^2/(8mL^2))(n+1)^2$, where E_C is the energy of the conduction band edge. Assuming $E_n^h = E_V - (h^2/(8mL^2))(n+1)^2$ for the holes (with E_V as the valence band edge) and a gap of $E_C - E_V = 1.4$ eV, how should the width L be chosen if we want emission in the red with $\lambda = 700$ nm? Can such an LED be realized technically?*

With the help of Figure 9.2, it follows that $L = 1.4$ nm = 14 Å, which is at the limit of the present possibilities of the technology in mass production.

Exercise 9.3 *An element of a detector array has to react for the wavelength 3000 nm. Assuming a quantum well width of 10 Å, and that Eq. (9.33) can be applied for the ground state, how deep should the well be?*

$V = 0.79$ eV.

Exercise 9.4 *Calculate the expectation value of the energy in the ground state of a quantum well, making use of the fact that the bottom of the well is at $V=0$, so the energy is purely kinetic and determined by p^2. Compare the result with the energy of the zero point vibration in Eq. (9.33)! (Hint: $\int_{-L/2}^{+L/2} \cos^2\left(\frac{\pi}{L}x\right) dx = \frac{L}{2}$.)*

In the ground state $\varphi_0(x) = \sqrt{(2/L)}\cos(\pi(x/L))$, it is true that $\langle p^2\rangle/2m = (\hbar^2\pi^2)/(2mL^2)$, which is equal to E_0.

Exercise 9.5 *Try to derive Eq. (9.33) from the interference condition for a string of length L! Use Eq. (8.10) for the momentum.*

$$E_n = \frac{p_n^2}{2m} = \frac{1}{2m}\frac{h^2}{\lambda_n^2} = \frac{1}{2m}\frac{h^2}{\lambda_n^2} = \frac{1}{2m}\frac{h^2(n+1^2)}{4L^2}$$

Chapter 10

Exercise 10.1 *An electron of energy 3 eV is approaching a potential wall of height $V_0 = 3.5$ eV and width 5 Å. The potential outside the wall is zero.*

(a) *What is the tunneling rate for an electron ($m_e = 9.10939 \times 10^{-31}$ kg)? Compare the exponential approximation and the exact formula of Eq. (10.10).*
$T_{\text{exact}} = 0.052354$, $T_{\text{exponential}} = 0.052335$.

(b) *Which formula would you choose for a proton ($m_p = 1.67263\times 10^{-27}$ kg) of the same energy? What is the tunneling rate for the proton?*
$T_{\text{exact}} = 7.519496 \times 10^{-68} = T_{\text{exponential}}$.

Exercise 10.2 *The band gaps of diamond and silicon are $E_g^{\text{Diam}} = 5.34$ eV and $E_g^{\text{Si}} = 1.12$ eV, respectively. Estimate the relation of the electric breakdown fields in the two materials. N.B.: comparison of the result with the experimentally observed ratio, 100/6, indicates that the effective mass of electrons (due to interaction between the wave-like electrons), is different in the two materials (see Section 5.8 and Chapter 12).*

Assuming the same leakage current in the two materials at the breakdown field strength $|\mathbf{E}_D|$, it follows that $|\mathbf{E}_D^{\text{Diam}}|/|\mathbf{E}_D^{\text{Si}}| = (E_g^{\text{Diam}}/E_g^{\text{Si}})^{3/2} = 10.41$.

Exercise 10.3 *The tunnel effect is being also utilized in flash memories. These are metal–oxide–semiconductor (MOS) structures, with an extra, so-called floating gate (FG) in the insulating layer (Figure 10.14). Information is stored by the charge state of the FG electrode (1: charged, 0: uncharged). Writing and erasing occurs with the help of the (original) control electrode (CG). Try to explain the role of the tunnel effect in this procedure, taking into account that the thickness of the insulating layer between the semiconductor and the FG is much smaller than between the FG and the CG.*

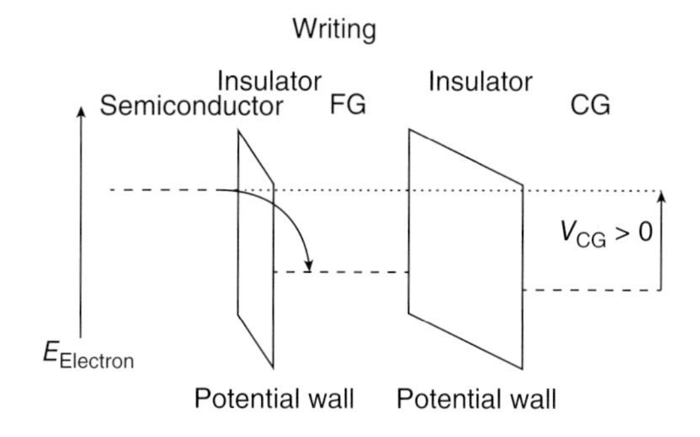

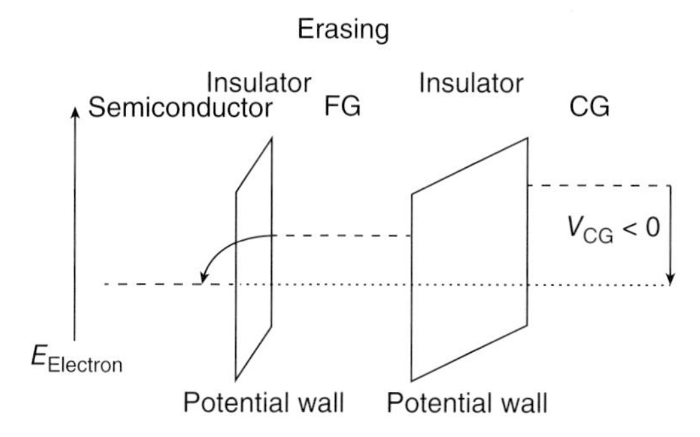

Since the second potential wall is thick, tunneling is only significant through the first one.

Exercise 10.4 *A quantum cascade laser contains five GaAs layers of thickness 2 nm each, separated by GaAlAs barriers (six layers) of width 1 nm each, as shown idealized in Figure 10.11.*

(a) *Light emission occurs by the transition of the electron from the first excited state into the ground state of the wells, after it has tunneled through the separating barrier. Which wavelength will be emitted? (Use the formula for infinite quantum wells.)*
$\lambda = 4396.25$ nm.

(b) *What is the necessary bias between the two sides of the whole sandwich structure? Assume a resonant tunneling rate of 100%. (Hint: the constant field strength is given by $|\mathbf{E}| = U/l$.)*
The energy must be shifted by 0.282 eV per 3 nm. Therefore, $U = 1.5$ V.

Chapter 11

Exercise 11.1 *The stationary states of the electron in the hydrogen atom are characterized by four quantum numbers.*

(a) *Which observables are provided by these quantum numbers and how?*

The principal quantum number n determines the energy $E_n = -(m_e q^4/8\varepsilon_0^2 h^2)(1/n^2)$

The angular momentum quantum number l determines the absolute value of the angular momentum $|\mathbf{L}_l| = \hbar\sqrt{l(l+1)}$.

The magnetic quantum number m determines the z-component of the angular momentum $L_{zm} = \hbar m$ (and, thereby, also the magnetic momentum $m_z = qL_z/(2m_e)$).

The spin quantum number s determines the z-component of the spin $S_z = (\hbar/2)s$.

(b) *What are the allowed values of the quantum numbers?*

$n = 1,2,\ldots; l = 0,1,\ldots, n-1; m = -l, \ldots, +l; s = \pm 1$.

Exercise 11.2 *Consider the electron states in the hydrogen atom with energy −0.8503 eV.*

(a) *What is the value of the corresponding main quantum number?*

$n = 4$

(b) *How many states can have this energy? Set up a table for the possible quantum numbers!*

32 (including spin).

(c) *What is the difference between these states of the same energy?*

The absolute value of the angular momentum and the orientation of the angular momentum and the spin.

(d) *Provide a general formula for the polar angle (to the z-axis) of the angular momentum! Can this angle be zero?*

$\cos\theta = m/\sqrt{l(l+1)}$, which leads to $\theta = 0$ only for $l = \infty$ (i.e., $E = \infty$).

(e) *Assuming all possible states (including spin) to be occupied by an electron, calculate the sum of the z-components of the total angular momentum!*

Since $\sum S_z = -\frac{\hbar}{2} + \frac{\hbar}{2} = 0$ for every m and $\sum_{m=-l}^{l} L_{zm} = \sum_{m=-l}^{l} \hbar m = 0$ for every l,

$$\sum_{nlms} J_z^{nmls} = 0.$$

Exercise 11.3 *Which wavelength is radiated at the electron transition from orbital m, depicted in Figure 11.8, to the orbital n? The ionization energy of the hydrogen atom is −13.605 eV.*

$$\lambda = \frac{c}{\nu} = \frac{hc}{\Delta E} = hc\left[\frac{m_e q^4}{8\varepsilon_0^2 h^2}\left(\frac{1}{3^2} - \frac{1}{4^2}\right)\right]^{-1} = 1874.7\,\text{nm}$$

Exercise 11.4 *What does quantum mechanics say about the atomic orbitals and how does that differ from the Bohr model?*

The probability density for finding the electron is described by the square of the orbital functions 1s, 2s, $2p_x$, $2p_y$, $2p_z$, and so on. The electron is neither localized in one point nor along the perimeter of a circle (although, for the highest l

value at a given n, the *radial* probability density is maximal at the corresponding Bohr radius r_n). Energy and angular momentum are quantized. The orbital angular momentum in the ground state is zero.

Chapter 12

Exercise 12.1 *The z-component of the photon spin is ±ħ. What kind of a symmetry property must a multi-photon wave function have?*
$\Psi\left(\ldots \mathbf{r}_i \ldots \mathbf{r}_j \ldots\right) = \Psi\left(\ldots \mathbf{r}_j \ldots \mathbf{r}_i \ldots\right)$

Exercise 12.2 *What is the energy of the ground state of eight noninteracting electrons in a one-dimensional, infinite quantum well of width L (with V = 0 at the bottom of the well)? According to you expectation, how would the energy change if interactions between the electrons are not neglected?*
$E_{tot} = 22.56$ eV and the interaction would increase it.

Exercise 12.3 *The valence electrons of a semiconductor can be considered, to a first approximation, as being trapped in a very wide quantum well. If applying the one-electron approximation, how are the interactions between the electrons taken into account? The fictitious independent one-electrons can be considered to be the charge carriers in the electric current. What consequences can the hidden quantum mechanical interactions have on their properties?*
The one-electron wave functions are determined by the pseudo-Schrödinger equation, derived from the variational principle and containing the Coulomb and exchange interaction between the one-electrons. A consequence of this is that these one-electron states cannot be characterized by the free electron mass. The "built-in" interactions influence the inertia against the effect of an external force.

Exercise 12.4 *Try to explain why the one-electron energies form a band in metals (cf. Figure 3.2)!*
Due to the Pauli principle, the huge number of identical atomic states in a crystal cannot have the same energy, so the interaction shifts the atomic levels relative to each other. This gives rise to a quasi-continuous distribution of energy levels within an energy band.

Exercise 12.5 *One cannot generate really random numbers by using traditional logic gates, while quantum computers can very well do that. Try to explain that, considering what can be expected when measuring (reading) a qubit, which is in the superposition state* $\frac{1}{\sqrt{2}}(|0\rangle + |1\rangle)$*!*
In this mixed states, the eigenfunctions have the same coefficients. Therefore, the probability to observe either of the two is fifty-fifty.

List of Figures

Essential Quantum Mechanics for Electrical Engineers, First Edition. Peter Deák.

Index

Essential Quantum Mechanics for Electrical Engineers, First Edition. Peter Deák.